CAMBRIDGE TRACTS IN MATHEMATICS

68. *Gibbs states on countable sets*

TO CECILIA

CHRISTOPHER J. PRESTON
Fellow of Lincoln College, Oxford

Gibbs states on countable sets

CAMBRIDGE UNIVERSITY PRESS

Published by the Syndics of the Cambridge University Press
Bentley House, 200 Euston Road, London NW1 2DB
American Branch: 32 East 57th Street, New York, N.Y.10022

Library of Congress Catalogue Card Number: 73-88310

ISBN: 0 521 203759

First published 1974

Printed in Great Britain
at the University Printing House, Cambridge
(Brooke Crutchley, University Printer)

Contents

Preface

In the last few years there has been a great deal of interest in problems arising in classical lattice statistical mechanics. The aim of this book is to provide to mathematicians with no background in physics an introduction to some of the results in this field. As the average mathematician probably has difficulty in understanding the language of mathematical physics, the approach of the book is to consider the subject as a branch of probability theory. It is thus assumed that the reader is acquainted with some of the basic facts of probability theory (e.g. σ-algebras, probability measures, finite state Markov chains), but apart from this the material is self-contained.

The basic objects to be studied will be certain classes of probability measures on $\mathscr{P}(S)$, where S is a set (finite or countably infinite) and $\mathscr{P}(S)$ denotes the set of subsets of S. The points of S can be interpreted as sites, each of which can be either empty or occupied by a particle, and the subset of $A \in \mathscr{P}(S)$ can be regarded as denoting when there are particles at exactly the points in A. Thus the probability measures on $\mathscr{P}(S)$ describe the distribution of configurations of particles; and they will usually represent the equilibrium distribution of some physical model.

There are three parts to the book: the first part is Chapters 1, 2 and 3, and in this the points of S are the vertices of a finite graph; the second part is Chapter 4, where the points of S are the vertices of a countable graph; the rest of the book constitutes the third part, in which the set S has no additional structure.

The first two parts deal with models where the interaction of the particles has some connection with the graph structure, namely the interaction is only between particles occupying sites of the graph that are nearest neighbours. This leads to a class of measures on $\mathscr{P}(S)$ called Markov random fields. It is shown in the first two parts that this class is the same as another class of measures (which arise in statistical mechanics), the Gibbs states given by nearest neighbour potentials.

The third part also deals with Gibbs states, in this case with the graph structure removed. In this part models are considered with the following property: if Λ is a finite subset of S, $A \subset \Lambda$ and $X \subset S - \Lambda$, then the conditional probability of there being particles on Λ at exactly the points of A, given that on $S - \Lambda$ there are particles at exactly the points of X, is specified. (This says that if we know what is happening outside a finite subset of S then we can compute the distribution of particles inside the finite set.) Denote the above conditional probability by $f^{\Lambda}(A, X)$. The relations which the $f^{\Lambda}(A, X)$ must satisfy are determined in Chapter 5, and in the following chapters an attempt is made to find out under what conditions the $f^{\Lambda}(A, X)$ do or do not uniquely determine a probability measure on $\mathscr{P}(S)$. This possible non-uniqueness corresponds in the language of statistical mechanics to the phenomenon of phase transition.

There is a slight duplication between the second and third parts in order to make the third part independent of the rest of the material (and thus it is possible to start reading the book at Chapter 5). At the end of most of the chapters there are some notes which refer to the Bibliography at the end of the book. The Bibliography is by no means complete but should serve as an introduction for someone new to the subject; for further new results the reader should look at the current issues of journals like *Communications in Mathematical Physics* and *The Journal of Mathematical Physics*.

This book grew out of some notes written in the summer of 1972. I am deeply indebted to Frank Spitzer who introduced me to statistical mechanics and taught me a lot of the ideas and techniques used in the book, and to Adriano Garsia for teaching me probability, and to both for advice and encouragement. Thanks also to John Kingman and Geoffery Grimmett for reading the manuscript and for their helpful comments. The work on the material in the book was done between July 1971 and January 1973, and the author acknowledges support from Air Force Grant AF-AFOSR-2088 at the University of California, San Diego, and also a Science Research Council Fellowship and an IBM Research Fellowship at the Mathematical Institute, Oxford University.

Lincoln College, Oxford
May 1973

CHRISTOPHER J. PRESTON

1. *Gibbs states and Markov random fields*

Let Λ be a finite set and let $\mathscr{P}(\Lambda)$ denote the set of subsets of Λ. In this chapter we will look at various classes of probability measures on $\mathscr{P}(\Lambda)$ that might arise in simple physical and biological models. The points of Λ can be interpreted as sites, each of which can be either empty or occupied by a particle (or some other entity); the subset $A \in \mathscr{P}(\Lambda)$ will be regarded as describing the state of the model when the points of A are occupied and the points of $\Lambda - A$ are empty. The elements of $\mathscr{P}(\Lambda)$ will sometimes be called *configurations*. Most physical models (even simple ones) involving configurations would be dynamic in nature; the probability measures that we will look at will describe the distribution of the configurations when the model is in some state of dynamic equilibrium.

The set Λ, representing the sites in the model, can be expected to have some additional structure, for example we might know the distances between the sites, or we might know that certain sites are connected. We will consider structures on Λ of the latter kind, thus we suppose that the points of Λ are the vertices of some finite graph $\mathscr{G} = (\Lambda, e)$, where e is the set of edges of $\mathscr{G}$. We do not allow $\mathscr{G}$ to have any multiple edges or loops. The following notation and definitions will be used: If $x, y \in \Lambda$ and there is an edge of the graph between x and y then we will say that x and y are *neighbours*. Let $c: \Lambda \times \Lambda \to \{0, 1\}$ be given by

$$c(x,y) = \begin{cases} 1 & \text{if } x \text{ and } y \text{ are neighbours,} \\ 0 & \text{otherwise.} \end{cases}$$

(Note that $c(x,x) = 0$ for all $x \in \Lambda$, since we did not allow $\mathscr{G}$ to have any loops.) If $A \in \mathscr{P}(\Lambda)$ then we define $\partial A \in \mathscr{P}(\Lambda)$ by

$$\partial A = \{y \in \Lambda - A : c(x,y) = 1 \quad \text{for some} \quad x \in A\}.$$

∂A is called the *boundary of* A; if $x \in \Lambda$ then we will write ∂x instead of $\partial\{x\}$ (and in fact we will in general tend to write x instead of $\{x\}$, thus for example if $A \subset \Lambda$ then we will write $A \cup x$ instead of $A \cup \{x\}$). For any $A \in \mathscr{P}(\Lambda)$, $|A|$ will denote the number of points in A; a subset $B \in \mathscr{P}(\Lambda)$ will be called a *simplex* of the graph if $B \neq \varnothing$ and if given any $x, y \in B$ with $x \neq y$ then $c(x,y) = 1$. We will say that $B \in \mathscr{P}(\Lambda)$ is an *n-simplex* (for $n \geqslant 0$) if B is a simplex and $|B| = n+1$. Note that for all $x \in \Lambda$ $\{x\}$ is a simplex of the graph.

We will let $\mathscr{S}(\Lambda)$ denote the set of probability measures on $\mathscr{P}(\Lambda)$ (with $\mathscr{P}(\Lambda)$ given the σ-algebra $\mathscr{P}(\mathscr{P}(\Lambda))$, i.e. the σ-algebra of all subsets of $\mathscr{P}(\Lambda)$). An element $\mu \in \mathscr{S}(\Lambda)$ describes the distribution of configurations when some model is in dynamic equilibrium; because of this we will call $\mathscr{S}(\Lambda)$ the set of *states* on Λ. Since $\mathscr{P}(\Lambda)$ is a finite set we will identify $\mu \in \mathscr{S}(\Lambda)$ with its density; thus we consider μ as a function from $\mathscr{P}(\Lambda)$ to $\mathbb{R}$ (where $\mathbb{R}$ denotes the real numbers) with the properties $\mu(A) \geqslant 0$ for all $A \in \mathscr{P}(\Lambda)$, and

$$\sum_{A \subset \Lambda} \mu(A) = 1.$$

The first class of states that we consider are the Gibbs states, which arise in models in statistical physics. A function

$$V: \mathscr{P}(\Lambda) \to \mathbb{R}$$

will be called a *potential* on Λ if $V(\varnothing) = 0$. For any potential V on Λ we define the *Gibbs state with potential* V to be the state π on Λ given by

$$\pi(A) = Z^{-1} \exp V(A) \quad \text{for all} \quad A \in \mathscr{P}(\Lambda),$$

where Z is the correct normalizing constant, i.e.

$$Z = \sum_{B \subset \Lambda} \exp V(B).$$

If V is a potential on Λ then we define $J_V: \mathscr{P}(\Lambda) \to \mathbb{R}$ by

$$J_V(A) = \sum_{X \subset A} (-1)^{|A-X|} V(X).$$

Then for any $A \in \mathscr{P}(\Lambda)$ we have

$$V(A) = \sum_{B \subset A} J_V(B).$$

(This follows immediately from what was once called the inclusion–exclusion principle and now goes under the name of the Möbius inversion formula; in any case it is a simple matter to check directly that it holds.) We will call J_V the *interaction potential* corresponding to V. Conversely, given a potential Φ let us denote by U_Φ the potential given by

$$U_\Phi(A) = \sum_{B \subset A} \Phi(B).$$

Thus Φ is the interaction potential corresponding to U_Φ.

It should be noted that if μ is any state on Λ with positive density (i.e. $\mu(A) > 0$ for all $A \subset \Lambda$) then μ is the Gibbs state with potential V where V is the potential given by

$$V(A) = \log\left[\frac{\mu(A)}{\mu(\varnothing)}\right] \quad \text{for any} \quad A \subset \Lambda.$$

Thus the class of Gibbs states consists of all states on Λ with positive density, and hence the introduction of potentials should be regarded as merely a convenient way of describing the states with positive density. The potentials that we will be most interested in, in this chapter, are those that have some connection with the graph structure: a potential V will be called a *nearest neighbour potential* if $J_V(A) \neq 0$ only when A is a simplex of the graph.

The second class of states on Λ that we look at are the Markov random fields. If $\mu \in \mathscr{S}(\Lambda)$ then we will say that μ is a *Markov random field* if:

(i) μ has positive density, i.e. $\mu(A) > 0$ for all $A \subset \Lambda$.

(ii) Given $x \notin A \subset \Lambda$ then the conditional probability (with respect to μ) that a configuration contains x, given that the configuration is A on $\Lambda - x$, is the same as the conditional probability that the configuration contains x given that the configuration is $A \cap \partial x$ on ∂x; i.e.

$$\frac{\mu(A \cup x)}{\mu(A \cup x)+\mu(A)} = \sum_{B \subset \Lambda-(\partial x \cup x)} \mu((A \cap \partial x) \cup x \cup B) \Big/ \sum_{B \subset \Lambda-(x \cup x)} [\mu((A \cap \partial x) \cup x \cup B) + \mu((A \cap \partial x) \cup B)].$$

This says that the probability of there being a particle at x, given a particular configuration of particles on $\Lambda - x$, only depends on what happens on the neighbours of x; thus in some sense particles do not interact unless they occupy neighbouring sites.

From the above description of a Markov random field one might expect a connection between Markov random fields and Gibbs states whose potentials have some connection with the graph structure. In fact the main point of this chapter is to show that Markov random fields and Gibbs states with nearest neighbour potentials are the same.

The above definition of a Markov random field is clearly quite cumbersome to work with, so we will define another class of states, called nearest neighbour states, which have a simpler definition and then show that they are the same as Markov random fields. Let $\mu \in \mathscr{S}(\Lambda)$; we will call μ a *nearest neighbour state* if:

(i) $\mu(A) > 0$ for all $A \subset \Lambda$.

(ii) Given $x \notin A \subset \Lambda$ then

$$\frac{\mu(A \cup x)}{\mu(A)} = \frac{\mu((A \cap \partial x) \cup x)}{\mu(A \cap \partial x)}.$$

Note that (ii) is equivalent to

$$\frac{\mu(A \cup x)}{\mu(A \cup x)+\mu(A)} = \frac{\mu((A \cap \partial x) \cup x)}{\mu((A \cap \partial x) \cup x)+\mu(A \cap \partial x)} \quad \text{for all} \quad x \notin A \subset \Lambda,$$

which says that the conditional probability that the configuration contains x, given that the configuration is A on $\Lambda - x$, is the same as the conditional probability that the configuration contains x given that the configuration is $A \cap \partial x$ on $\Lambda - x$.

PROPOSITION 1.1 If $\mu \in \mathscr{S}(\Lambda)$ then μ is a Markov random field if and only if it is a nearest neighbour state.

Proof Let μ be a Markov random field and let $x \notin A \subset \Lambda$. Then

$$\frac{\mu(A \cup x)}{\mu(A \cup x)+\mu(A)} = \sum_{B \subset \Lambda-(\partial x \cup x)} \mu((A \cap \partial x) \cup x \cup B)$$
$$\times \left[\sum_{B \subset \Lambda-(\partial x \cup x)} [\mu((A \cap \partial x) \cup x \cup B)+\mu((A \cap \partial x) \cup B)]\right]^{-1}$$
$$= \frac{\mu((A \cap \partial x) \cup x)}{\mu((A \cap \partial x) \cup x)+\mu(A \cap \partial x)},$$

since the middle line is unchanged on replacing A by $A \cap \partial x$. Thus μ is a nearest neighbour state. Conversely, suppose that μ is a nearest neighbour state and again let $x \notin A \subset \Lambda$. If

$$B \subset \Lambda-(\partial x \cup x)$$

then

$$\frac{\mu((A \cap \partial x) \cup B \cup x)}{\mu((A \cap \partial x) \cup B)} = \frac{\mu((((A \cap \partial x) \cup B) \cap \partial x) \cup x)}{\mu(((A \cap \partial x) \cup B) \cap \partial x)}$$
$$= \frac{\mu((A \cap \partial x) \cup x)}{\mu(A \cap \partial x)} = \frac{\mu(A \cup x)}{\mu(A)}.$$

Thus

$$\mu(A)\,\mu((A \cap \partial x) \cup B \cup x) = \mu(A \cup x)\,\mu((A \cap \partial x) \cup B)$$

and therefore

$$\mu(A) \sum_{B \subset \Lambda-(\partial x \cup x)} \mu((A \cap \partial x) \cup B \cup x)$$
$$= \mu(A \cup x) \sum_{B \subset \Lambda-(\partial x \cup x)} \mu((A \cap \partial x) \cup B).$$

From this it is easy to see that μ is a Markov random field. □

We will now show that Markov random fields and Gibbs states with nearest neighbour potentials are the same. We start by proving:

PROPOSITION 1.2 Let π be the Gibbs state with nearest neighbour potential V. Then π is a Markov random field.

Proof By Proposition 1.1 we need only show that π is a nearest neighbour state, and it is clear that π has positive density. If $x \notin A \subset \Lambda$ then we have

$$\frac{\pi(A \cup x)}{\pi(A)} = \exp\left[V(A \cup x) - V(A)\right].$$

But

$$V(A \cup x) - V(A) = \sum_{B \subset A \cup x} J_V(B) - \sum_{B \subset A} J_V(B)$$

$$= \sum_{B \subset A} J_V(B \cup x) = \sum_{B \subset A} J_V((B \cap \partial x) \cup x)$$

$$= V((A \cap \partial x) \cup x) - V(A \cap \partial x).$$

Therefore $$\frac{\pi(A \cap x)}{\pi(A)} = \frac{\pi((A \cap \partial x) \cup x)}{\pi(A \cap \partial x)},$$

and thus π is a nearest neighbour state. □

The proof of the converse of this result is just as easy:

Proposition 1.3 Let μ be a Markov random field. Then there exists a unique nearest neighbour potential V such that μ is the Gibbs state with potential V.

Proof It is clear that V is unique since we are forced to define V by

$$V(A) = \log\left[\frac{\mu(A)}{\mu(\varnothing)}\right] \quad \text{for all} \quad A \in \mathscr{P}(\Lambda).$$

Then we have that μ is the Gibbs state with potential V, so it only remains to prove that V is a nearest neighbour potential. Let $A \in \mathscr{P}(\Lambda)$ with A not a simplex of the graph. Then there exists $x, y \in A$ with $x \neq y$ and $c(x, y) = 0$. Put $B = A - x - y$; then if $X \subset B$ we have

$$\frac{\mu(X \cup x \cup y)}{\mu(X \cup x)} = \frac{\mu(((X \cup x) \cap \partial y) \cup y)}{\mu((X \cup x) \cap \partial y)}$$

$$= \frac{\mu((X \cap \partial y) \cup y)}{\mu(X \cap \partial y)} = \frac{\mu(X \cup y)}{\mu(X)};$$

thus $$V(X \cup x \cup y) - V(X \cup x) - V(X \cup y) + V(X) = 0.$$

Now we have

$$\begin{aligned}
J_V(A) &= \sum_{E \subset A} (-1)^{|A-E|} V(E) \\
&= \sum_{X \subset B} [(-1)^{|A-X\cup x\cup y|} V(X \cup x \cup y) + (-1)^{|A-X\cup x|} V(X \cup x) \\
&\qquad + (-1)^{|A-X\cup y|} V(X \cup y) + (-1)^{|A-X|} V(X)] \\
&= \sum_{X \subset B} (-1)^{|A-X|} [V(X \cup x \cup y) - V(X \cup x) - V(X \cup y) + V(X)] \\
&= 0,
\end{aligned}$$

and thus V is a nearest neighbour potential. □

For the sake of reference we now collect together our previous results and call them a theorem.

THEOREM 1.1 The following are equivalent for $\mu \in \mathscr{S}(\Lambda)$:
(i) μ is a Markov random field;
(ii) μ is a nearest neighbour state;
(iii) μ is a Gibbs state with potential V for some nearest neighbour potential V.

A re-examination of the proof of Proposition 1.3 suggests the following characterization of nearest neighbour potentials:

PROPOSITION 1.4 Let V be a potential on Λ. Then V is a nearest neighbour potential if and only if given $x, y \in \Lambda$ with $x \neq y$ and $c(x, y) = 0$ and given $X \subset \Lambda - x - y$ then

$$V(X \cup x \cup y) - V(X \cup x) - V(x \cup y) + V(X) = 0.$$

Proof If the condition holds, then the proof of Proposition 1.3 shows that V is a nearest neighbour potential. Conversely suppose that V is a nearest neighbour potential and let X, x, y be as in the hypothesis. If $B \subset X$ then just as in Proposition 1.3 we have

$$J_V(B \cup x \cup y) = \sum_{Y \subset B} (-1)^{|B-Y|} [V(Y \cup x \cup y) - V(Y \cup x) - V(Y \cup y) + V(Y)].$$

But $B \cup x \cup y$ is not a simplex of the graph since $c(x,y) = 0$, and hence $J_V(B \cup x \cup y) = 0$. Thus for all $B \subset X$ we have

$$\sum_{Y \subset B} (-1)^{|B-Y|}[V(Y \cup x \cup y) - V(Y \cup y) - V(Y \cup x) + V(Y)] = 0.$$

Therefore

$$\begin{aligned} &V(X \cup x \cup y) - V(X \cup x) - V(X \cup y) + V(X) \\ &\quad = \sum_{B \subset X} \sum_{Y \subset B} (-1)^{|B-Y|}[V(Y \cup x \cup y) - V(Y \cup x) - V(Y \cup y) + V(Y)] \\ &\quad = 0. \square \end{aligned}$$

We will use the (confusing) terminology of the physicists and say that the graph $\mathscr{G}$ is a *cubic lattice* if $\mathscr{G}$ contains no 2-simplexes (and thus contains no n-simplexes for $n \geqslant 2$). The most common occurrences of cubic lattices are as finite subsets of $\mathbb{Z}^\nu$ (where for $\nu \geqslant 1$, $\mathbb{Z}^\nu$ denotes the points of $\mathbb{R}^\nu$ which have integer co-ordinates, considered as a graph by defining two points to be neighbours if the distance between them is exactly 1). Let $\mathscr{G} = (\Lambda, e)$ be a cubic lattice and let V be a nearest neighbour potential on Λ. Then we must have $J_V(A) = 0$ if $|A| \geqslant 3$, and thus defining $H: \Lambda \times \Lambda \to \mathbb{R}$ by

$$H(x,y) = \begin{cases} \frac{1}{2} V(\{x,y\}) & \text{if} \quad x \neq y, \\ V(\{x\}) & \text{if} \quad x = y, \end{cases}$$

then it is easy to check that we have

$$V(A) = \sum_{x \in A} \sum_{y \in A} H(x,y) \quad \text{for all} \quad A \in \mathscr{P}(\Lambda).$$

It is also clear that $H(x,y) = 0$ if $x \neq y$ and $c(x,y) = 0$. This shows that for cubic lattices we can write nearest neighbour potentials in the form of nearest neighbour pair potentials as they are usually defined in elementary physics.

Notes The definition of a Gibbs state (on a finite subset of $\mathbb{Z}^\nu$) goes back to the classical work of Gibbs (1902). Markov random fields (on $\mathbb{Z}^\nu$) were first introduced by Dobrushin (1968a). Spitzer (1971a) and Averintsev (1970) showed that for finite subsets of $\mathbb{Z}^\nu$ the class of Markov random fields and Gibbs states with nearest neighbour potentials were the same (in Spitzer's paper translation invariance was assumed). The equivalence of Markov random fields and Gibbs states with nearest neighbour potentials on general finite graphs was first obtained by Hammersley and Clifford (1971). Their proof was long and involved and a considerably easier proof is given in Preston (1973), an even simpler proof is given in Grimmett (1973), and yet another proof is given in Sherman (1973). The proof given in this book is an adaptation of Grimmett's proof. For a different approach to this material see Suomela (1972).

2. *Interacting particle systems*

The aim of this chapter is to investigate some simple dynamic models that have the probability measures considered in the previous chapter as their equilibrium states. We will look at models in which the dynamics is random; in fact we will consider Markov chains. As before $\mathscr{G} = (\Lambda, e)$ will be an arbitrary finite graph, and for convenience we will denote $\mathscr{P}(\Lambda)$ by Γ.

Let $\{P_t\}_{t\geqslant 0}$ be a semi-group on Γ; i.e. for each $t \geqslant 0$ we have $P_t: \Gamma\times\Gamma \to \mathbb{R}$ with the properties:

(i) $0 \leqslant P_t(A, B)$ for all $A, B \in \Gamma$, $t \geqslant 0$;

(ii) $\sum_{B\in\Gamma} P_t(A, B) = 1$ for all $A \in \Gamma$, $t \geqslant 0$;

(iii) $\sum_{X\in\Gamma} P_t(A, X)\, P_s(X, B) = P_{t+s}(A, B)$ for all $A, B \in \Gamma$, $s, t \geqslant 0$;

(iv) $\lim_{t\to 0} P_t(A, B) = I(A, B)$ for all $A, B \in \Gamma$, where

$$I(A, B) = \begin{cases} 1 & \text{if} \quad A = B, \\ 0 & \text{otherwise.} \end{cases}$$

The semi-group $\{P_t\}_{t\geqslant 0}$ is interpreted as describing a model which has the property that if the model has configuration $A \in \Gamma$ at time s then the probability that it will have configuration $B \in \Gamma$ at time $s+t$ is $P_t(A, B)$.

It is a well-known result that there exists a unique function $G: \Gamma\times\Gamma \to \mathbb{R}$ satisfying:

(i) $G(A, B) \geqslant 0$ if $A, B \in \Gamma$ and $A \neq B$;

(ii) $\sum_{B\in\Gamma} G(A, B) = 0$ for all $A \in \Gamma$;

(iii) $P_t = \exp(tG)$.

(Here we consider P_t and G as $|\Gamma| \times |\Gamma|$ matrices and define

$$\exp(tG) = \sum_{n=0}^{\infty} \frac{t^n}{n!} G^n,$$

with G^n the product of the matrix G with itself n times, and $G^0 = I$.) G is called the *generator* of the semi-group. The converse of the above result holds, namely, if $G:\Gamma\times\Gamma\to\mathbb{R}$ satisfies (i) and (ii) then defining P_t by $P_t = \exp(tG)$ gives a semi-group $\{P_t\}_{t\geqslant 0}$.

The generator G of the semi-group $\{P_t\}_{t\geqslant 0}$ has the following property: if $A, B\in\Gamma$ with $A \neq B$ then the conditional probability that the model will change from having configuration A to having configuration B between times t and $t+\mathrm{d}t$, given that it has configuration A at time t, is $G(A,B)\,\mathrm{d}t+O(\mathrm{d}t^2)$. We will construct various models by defining generators, and the interpretation of the models can be made from the above fact.

A state $\pi\in\mathscr{S}(\Lambda)$ is called an *equilibrium state* for the semi-group $\{P_t\}_{t\geqslant 0}$ if

$$\sum_{A\in\Gamma}\pi(A)\,P_t(A,B) = \pi(B)\quad\text{for all}\quad B\in\Gamma,\, t\geqslant 0.$$

It is not difficult to check that π is an equilibrium state if and only if

$$\sum_{A\in\Gamma}\pi(A)\,G(A,B) = 0\quad\text{for all}\quad B\in\Gamma.$$

The generator G is said to be *irreducible* if given $A, B\in\Gamma$ with $A\neq B$ then there exist $E_1,\ldots,E_n\in\Gamma$ with $A = E_1$, $B = E_n$ and $G(E_1,E_2)\,G(E_2,E_3)\ldots G(E_{n-1},E_n) > 0$. It is a well-known result (part of the ergodic theorem for finite Markov chains) that if the generator G is irreducible then there exists a unique equilibrium state π, π has positive density, and that for any $\mu\in\mathscr{S}(\Lambda)$ we have

$$\lim_{t\to\infty}\sum_{A\in\Gamma}\mu(A)\,P_t(A,B) = \pi(B)\quad\text{for all}\quad B\in\Gamma.$$

One final definition from the theory of Markov chains: we will say that a semi-group with irreducible generator G and equilibrium state π is *time-reversible* if

$$\pi(A)\,G(A,B) = \pi(B)\,G(B,A)\quad\text{for all}\quad A,B\in\Gamma.$$

(Note that any state π satisfying this condition must be an equilibrium state and thus *the* equilibrium state.) The condition of time-reversibility is such that if a film were made of the time evolution of the model, with the model started in its equilibrium state, then it would not be possible to distinguish between the film run forwards and the film run backwards.

We will now describe a model, which could be called a birth–death process on Λ. The configuration $A \in \Gamma$ describes when the model has a particle at each of the points of A, and does not have particles at the points of $\Lambda - A$. The model evolves by particles being born or dying. We are given two functions $\beta: \Lambda \times \Gamma \to \mathbb{R}$, $\delta: \Lambda \times \Gamma \to \mathbb{R}$, with $\beta(x, A) > 0$, $\delta(x, A) > 0$, and we suppose that the probability of a particle being born at x between times t and $t + \mathrm{d}t$, given that the configuration at time t is A (with $x \notin A \in \Gamma$), is $\beta(x, A)\,\mathrm{d}t + O(\mathrm{d}t^2)$. Similarly we suppose that the probability of a particle dying at x between times t and $t + \mathrm{d}t$, given that the configuration at time t is $A \cup x$ (with $x \notin A \in \Gamma$), is $\delta(x, A)\,\mathrm{d}t + O(\mathrm{d}t^2)$. Finally we assume that the probability of more than one birth or death between times t and $t + \mathrm{d}t$ is $O(\mathrm{d}t^2)$. We will call β the *birth rate* of the process, and δ the *death rate* of the process. The above suggests that if $\{P_t\}_{t \geqslant 0}$ is the semi-group describing the model then its generator G should be defined as follows:

$$G(A, A \cup x) = \beta(x, A) \quad \text{if} \quad x \notin A \in \Gamma,$$

$$G(A \cup x, A) = \delta(x, A) \quad \text{if} \quad x \notin A \in \Gamma,$$

$$G(A, B) = 0 \quad \text{for all other pairs } A, B \in \Gamma \text{ with } A \neq B.$$

This leave us still to define $G(A, A)$, but since G is a generator $G(A, A)$ is determined by the condition

$$\sum_{B \in \Gamma} G(A, B) = 0.$$

We will call the semi-group corresponding to G a *birth–death* semi-group with birth rate β and death rate δ. It is clear that G is irreducible and thus the semi-group has a unique equilibrium state π. We will say that the semi-group is a *nearest neighbour* birth–death semi-group if

$$\beta(x, A) = \beta(x, A \cap \partial x), \quad \delta(x, A) = \delta(x, A \cap \partial x) \quad \text{for all} \quad x \notin A \in \Gamma.$$

PROPOSITION 2.1 Let $\{P_t\}_{t\geqslant 0}$ be a time reversible nearest neighbour birth–death semi-group, and let π be its equilibrium state. Then π is a Markov random field.

Proof By Theorem 1.1 we need only show that π is a nearest neighbour state, and since π is the equilibrium state of a semi-group with an irreducible generator we immediately have that $\pi(A) > 0$ for all $A \in \Gamma$. The condition that the semi-group is time reversible reduces in the present case to having

$$\pi(A)\,G(A, A \cup x) = \pi(A \cup x)\,G(A \cup x, A) \quad \text{for all} \quad x \notin A \in \Gamma;$$

which is the same as having

$$\frac{\pi(A \cup x)}{\pi(A)} = \frac{\beta(x, A)}{\delta(x, A)} \quad \text{for all} \quad x \notin A \in \Gamma.$$

But by hypothesis we have

$$\frac{\beta(x, A)}{\delta(x, A)} = \frac{\beta(x, A \cap \partial x)}{\delta(x, A \cap \partial x)} \quad \text{for all} \quad x \notin A \in \Gamma,$$

and thus π is a nearest neighbour state. □

The converse of Proposition 2.1 is easily seen to be true; namely every Markov random field is the equilibrium state of some time reversible nearest neighbour birth–death semi-group. In fact, if μ is a nearest neighbour state then the birth–death semi-group with birth rate $\beta(x, A) = \mu(A \cup x)/\mu(A)$ and death rate $\delta(x, A) = 1$ is time reversible and has μ as its equilibrium state.

The reason that Proposition 2.1 is true depends largely on the fact that the semi-group is time reversible, a condition which places severe restrictions on the form that the birth rate and the death rate can take. To emphasize this fact we will ignore the graph structure on Λ for a moment and show the following:

PROPOSITION 2.2 Let $\{P_t\}_{t\geqslant 0}$ be a birth–death semi-group on Λ with birth rate β and death rate δ. The following are equivalent:

(i) $\{P_t\}_{t\geqslant 0}$ is time reversible.

(ii) Define $\gamma: \Lambda \times \Gamma \to \mathbb{R}$ by $\gamma(x, A) = \beta(x, A)/\delta(x, A)$; then if $A \in \Gamma$, $x, y \in \Lambda - A$ with $x \neq y$ we must have

$$\gamma(x, A \cup y)\,\gamma(y, A) = \gamma(y, A \cup x)\,\gamma(x, A).$$

(iii) There exists a potential $V: \mathscr{P}(\Lambda) \to \mathbb{R}$ such that

$$\gamma(x, A) = \exp\,[V(A \cup x) - V(A)] \quad \text{for all} \quad x \notin A \in \Gamma.$$

Proof (i) ⇒ (ii): Let $\{P_t\}_{t \geq 0}$ have generator G and equilibrium state π, and let $A \in \Gamma$, $x, y \in \Lambda - A$ with $x \neq y$. Since $\{P_t\}_{t \geq 0}$ is time reversible we must have

$$\gamma(x, A \cup y) = \frac{\pi(A \cup y \cup x)}{\pi(A \cup y)}, \quad \gamma(y, A) = \frac{\pi(A \cup y)}{\pi(A)},$$

$$\gamma(y, A \cup x) = \frac{\pi(A \cup x \cup y)}{\pi(A \cup x)}, \quad \gamma(x, A) = \frac{\pi(A \cup x)}{\pi(A)},$$

and thus $\quad \gamma(x, A \cup y)\,\gamma(y, A) = \gamma(y, A \cup x)\,\gamma(x, A).$

(ii) ⇒ (iii): We define $V(A)$ by induction on $|A|$. Since V is a potential we have to have $V(\varnothing) = 0$. Now let $A \in \Gamma$ with $|A| > 0$, choose $x \in A$, and define

$$V(A) = V(A - x) + \log \gamma(x, A - x).$$

Condition (ii) is exactly that to ensure that $V(A)$ is well-defined, i.e. it does not matter how we choose $x \in A$. By construction we certainly have

$$\gamma(x, A) = \exp\,[V(A \cup x) - V(A)] \quad \text{for all} \quad x \notin A \in \Gamma.$$

(iii) ⇒ (i): Let μ be the Gibbs state with potential V. Then it is a simple matter to check that

$$\mu(A)\,G(A, B) = \mu(B)\,G(B, A) \quad \text{for all} \quad A, B \in \Gamma,$$

and thus we have that $\{P_t\}_{t \geq 0}$ is time-reversible and also that $\mu = \pi$. □

We now consider another model, this time a model for the interaction of m indistinguishable particles moving around on Λ (where m is fixed with $m < |\Lambda|$). We exclude multiple occupancy, i.e. there can be at most one particle at any point of Λ. Let $\Gamma_m = \{A \in \Gamma : |A| = m\}$; we will define a semi-group $\{P_t\}_{t \geq 0}$, this time having $P_t : \Gamma_m \times \Gamma_m \to \mathbb{R}$, by specifying its generator

$$G : \Gamma_m \times \Gamma_m \to \mathbb{R}.$$

We are given a function $d : \Lambda \times \Gamma \to \mathbb{R}$ with $d(x, A) > 0$ for all $x \notin A \in \Gamma$, and we are also given an irreducible, symmetric transition function $P : \Lambda \times \Lambda \to [0, 1]$ with $P(x, x) = 0$ for all $x \in \Lambda$. We suppose that if $x \notin A \in \Gamma$ then the probability that a particle at x will jump somewhere between times t and $t + \mathrm{d}t$, given that at time t there are particles at exactly the points of $A \cup x$, is $d(x, A)\,\mathrm{d}t + O(\mathrm{d}t^2)$, and given that the particle will jump somewhere then it will jump to y with probability $P(x, y)$. We also suppose that the probability of more than one jump occurring between times t and $t + \mathrm{d}t$ is $O(\mathrm{d}t^2)$. Therefore we are led to define a generator $G : \Gamma_m \times \Gamma_m \to \mathbb{R}$ as follows:

$$G(A \cup x, A \cup y) = d(x, A)\,P(x, y) \quad \text{if} \quad A \in \Gamma_{m-1}, \quad x, y \in \Lambda - A, \quad x \neq y;$$

$$G(A, B) = 0 \quad \text{for all other pairs} \quad A, B \in \Gamma_m \quad \text{with} \quad A \neq B.$$

Once again $G(A, A)$ is defined to give

$$\sum_{B \in \Gamma_m} G(A, B) = 0.$$

We will call the semi-group corresponding to G an *m-particle* semi-group; d is called the *speed function* and P the *jump matrix* of the semi-group. Note that G is irreducible (on Γ_m) and thus there exists a unique equilibrium state π (with π a probability measure on Γ_m). We will call the semi-group a *nearest neighbour* m-particle semi-group if $d(x, A) = d(x, A \cap \partial x)$ for all $x \notin A \in \Gamma$.

PROPOSITION 2.3 Let $\{P_t\}_{t \geq 0}$ be an m-particle semi-group with speed function d. Suppose that if $A \in \Gamma$, $x, y \in \Lambda - A$ with $x \neq y$ then

$$d(x, A \cup y)\,d(y, A) = d(y, A \cup x)\,d(x, A).$$

Then we have:

(i) There exists a unique potential $V:\mathscr{P}(\Lambda)\to\mathbb{R}$ such that

$$d(x,A) = \exp[V(A\cup x) - V(A)] \quad \text{for all} \quad x\notin A\in\Gamma.$$

(ii) If π is the equilibrium state of $\{P_t\}_{t\geqslant 0}$ then

$$\pi(A) = Z^{-1}\exp V(A) \quad \text{for all} \quad A\in\Gamma_m,$$

where
$$Z = \sum_{B\in\Gamma_m} \exp V(B).$$

(iii) $\{P_t\}_{t\geqslant 0}$ is time-reversible.

(iv) If $\{P_t\}_{t\geqslant 0}$ is a nearest neighbour m-particle semi-group then V is a nearest neighbour potential.

Proof The proof of (i) is exactly the same as the proof of Proposition 2.2, (ii) and (iii) follow by direct calculation, and (iv) follows immediately from Proposition 1.4. □

As a partial converse to Proposition 2.3 we have:

PROPOSITION 2.4 Suppose that whenever $x,y\in\Lambda$ with $c(x,y) = 1$ then $|\partial x\cup\partial y|\leqslant m$ and $|\Lambda| - |\partial x\cup\partial y|\geqslant m-1$. Let π be the equilibrium state of a time-reversible, nearest neighbour m-particle semi-group. Then there exists a nearest neighbour potential V such that

$$\pi(A) = Z^{-1}\exp V(A) \quad \text{for all} \quad A\in\Gamma_m,$$

with
$$Z = \sum_{B\in\Gamma_m} \exp V(B).$$

(Thus π is the Gibbs state with potential V conditioned on the subset Γ_m of Γ.)

Proof By Proposition 2.3 it is sufficient to show that if d is the speed function of the semi-group and if $A\in\Gamma$, $x,y\in\Lambda - A$ with $x\neq y$, then

$$d(x,A\cup y)\,d(y,A) = d(y,A\cup x)\,d(x,A).$$

To prove that this holds it is sufficient to restrict ourselves to the case when $c(x,y) = 1$ and $A\subset(\partial x\cup\partial y)$ (since

$$d(z,B) = d(z,B\cap\partial z)$$

for all $z \notin B \in \Gamma$). Thus by hypothesis we have

$$|A| \leqslant m-2 \quad \text{and} \quad |\Lambda| - |\partial x \cup \partial y| \geqslant m-1,$$

so we can choose $E \subset \Lambda - (\partial x \cup \partial y)$ with $|E| = m-1-|A| > 0$. Let $B = A \cup E$, thus $|B| = m-1$, and choose $z \in E$. Now we have

$$\frac{d(x, A \cup y)}{d(y, A \cup x)} = \frac{d(x, B \cup y - z)}{d(y, B \cup x - z)}$$

$$= \frac{d(x, B \cup y - z)}{d(z, B \cup y - z)} \frac{d(z, B \cup y - z)}{d(z, B \cup x - z)} \frac{d(z, B \cup x - z)}{d(y, B \cup x - z)}$$

$$= \frac{\pi((B \cup y - z) \cup z)}{\pi((B \cup y - z) \cup x)} \frac{d(z, B \cup y - z)}{d(z, B \cup x - z)} \frac{\pi((B \cup x - z) \cup y)}{\pi((B \cup x - z) \cup z)}$$

(by time-reversibility, since $|B \cup y - z| = m-1$)

$$= \frac{\pi(B \cup y)}{\pi(B \cup x)} \frac{d(z, B \cup y - z)}{d(z, B \cup x - z)} = \frac{\pi(B \cup y)}{\pi(B \cup x)}$$

(since $(B \cup y - z) \cap \partial z = (B \cup x - z) \cap \partial z$)

$$= \frac{d(x, B)}{d(y, B)} = \frac{d(x, A)}{d(y, A)}$$

(by time-reversibility, since $|B| = m-1$, and also since

$$B \cap \partial x = A \cap \partial x, \quad B \cap \partial y = A \cap \partial y).$$

Therefore we do have

$$d(x, A \cup y)\, d(y, A) = d(y, A \cup x)\, d(x, A). \square$$

We should perhaps reinterpret the above result. What it really says is that an m-particle semi-group can only be a time-reversible nearest neighbour m-particle semi-group if its speed function d is of a very special form, namely we must have

$$d(x, A) = \exp[V(A \cup x) - V(A)] \quad \text{for all} \quad x \notin A \in \Gamma,$$

for some nearest neighbour potential V.

NOTES The fact that Markov random fields are the same as the equilibrium states of the Markov chains appearing in this chapter was first proved by Spitzer (1971b) in the case of a finite subset of $\mathbb{Z}^\nu$ and assuming translation invariance. The result for a general finite graph is given in Preston (1973). For generalizations of this chapter to countable graphs see the notes for Chapter 4. The standard material on Markov chains and ergodic theory can be found for example in Doob (1953) or Billingsley (1965).

3. *Coupled Markov chains*

The object of this chapter is to prove a theorem of Holley about coupled Markov chains (although the statement of the theorem does not appear to have anything to do with Markov chains) and to examine some consequences of the theorem. The theorem is:

THEOREM 3.1 Let Λ be a finite set and let μ_1, μ_2 be positive elements of $\mathscr{S}(\Lambda)$ (i.e. we have $\mu_1(A) > 0$, $\mu_2(A) > 0$ for all $A \in \mathscr{P}(\Lambda)$). Suppose that for all $A, B \in \mathscr{P}(\Lambda)$

$$\mu_1(A \cup B)\,\mu_2(A \cap B) \geqslant \mu_1(A)\,\mu_2(B).$$

Then there exists a probability measure ν on $\mathscr{P}(\Lambda) \times \mathscr{P}(\Lambda)$ such that

(i) $\sum_{B \subset \Lambda} \nu(A, B) = \mu_1(A)$ for all $A \subset \Lambda$.

(ii) $\sum_{A \subset \Lambda} \nu(A, B) = \mu_2(B)$ for all $B \subset \Lambda$.

(iii) $\nu(A, B) = 0$ unless $A \supset B$.

Let us accept for the moment that the theorem is true, then from it we can deduce:

THEOREM 3.2 Let Λ be a finite set and let μ_1, μ_2 be positive elements of $\mathscr{S}(\Lambda)$ such that for all $A, B \in \mathscr{P}(\Lambda)$ we have

$$\mu_1(A \cup B)\,\mu_2(A \cap B) \geqslant \mu_1(A)\,\mu_2(B).$$

Let $f: \mathscr{P}(\Lambda) \to \mathbb{R}$ be increasing (i.e. if $A \supset B$ then $f(A) \geqslant f(B)$). Then

$$\sum_{A \subset \Lambda} f(A)\,\mu_1(A) \geqslant \sum_{A \subset \Lambda} f(A)\,\mu_2(A).$$

Proof Let ν be the probability measure on $\mathscr{P}(\Lambda) \times \mathscr{P}(\Lambda)$ defined in Theorem 3.1. Then we have

$$\begin{aligned}\sum_{A \subset \Lambda} f(A)\mu_1(A) &= \sum_{A \subset \Lambda} \sum_{B \subset \Lambda} f(A)\,\nu(A,B) \\ &= \sum_{A \subset \Lambda} \sum_{B \subset A} f(A)\,\nu(A,B) \geqslant \sum_{A \subset \Lambda} \sum_{B \subset A} f(B)\,\nu(A,B) \\ &= \sum_{A \subset \Lambda} \sum_{B \subset \Lambda} f(B)\,\nu(A,B) = \sum_{B \subset \Lambda} f(B)\,\mu_2(B). \square\end{aligned}$$

We will call the inequality given in Theorem 3.2 *Holley's inequality*. It will be of fundamental importance for obtaining many results in Chapter 8. From Holley's inequality follows another inequality, discovered by Fortuin, Kastelyn and Ginibre (and thus called the *FKG inequality*) which is given as follows:

PROPOSITION 3.1 Let Λ be a finite set and let μ be a positive element of $\mathscr{S}(\Lambda)$ such that for all $A, B \in \mathscr{P}(\Lambda)$ we have

$$\mu(A \cup B)\,\mu(A \cap B) \geqslant \mu(A)\,\mu(B).$$

Let $f, g: \mathscr{P}(\Lambda) \to \mathbb{R}$ be increasing. Then

$$\sum_{A \subset \Lambda} f(A)\,g(A)\,\mu(A) \geqslant \sum_{A \subset \Lambda} f(A)\,\mu(A) \sum_{B \subset \Lambda} g(B)\,\mu(B).$$

Proof If $\alpha \in \mathbb{R}$ then it is clear that each side of the inequality changes by the same amount on replacing g by $g+\alpha$; thus without loss of generality we can assume that $g(A) > 0$ for all $A \in \mathscr{P}(\Lambda)$. Define $\mu_1, \mu_2 \in \mathscr{S}(\Lambda)$ by $\mu_2 = \mu$, and

$$\mu_1(A) = Z^{-1} g(A)\,\mu(A) \quad \text{for all} \quad A \in \mathscr{P}(\Lambda),$$

with of course
$$Z = \sum_{B \subset \Lambda} g(B)\,\mu(B).$$

Then if $A, B \in \mathscr{P}(\Lambda)$ we have

$$\begin{aligned}\mu_1(A \cup B)\,\mu_2(A \cap B) &= Z^{-1} g(A \cup B)\,\mu(A \cup B)\,\mu(A \cap B) \\ &\geqslant Z^{-1} g(A)\,\mu(A)\,\mu(B) = \mu_1(A)\,\mu_2(B),\end{aligned}$$

and thus by Holley's inequality

$$\sum_{A\subset\Lambda} f(A)\,g(A)\,\mu(A) = Z\sum_{A\subset\Lambda} f(A)\,\mu_1(A) \geqslant Z\sum_{A\subset\Lambda} f(A)\,\mu_2(A)$$

$$= \sum_{A\subset\Lambda} f(A)\,\mu(A)\sum_{B\subset\Lambda} g(B)\,\mu(B).\square$$

We will now start the proof of Theorem 3.1. Let μ be a positive element of $\mathscr{S}(\Lambda)$ and define a function $d:\Lambda\times\mathscr{P}(\Lambda)\to\mathbb{R}$ by

$$d(x,A) = \frac{\mu(A-x)}{\mu(A)}.$$

Also let us define $G:\mathscr{P}(\Lambda)\times\mathscr{P}(\Lambda)\to\mathbb{R}$ by

$$G(A,A-x) = d(x,A) \quad \text{if} \quad x\in A\subset\Lambda;$$

$$G(A,A\cup x) = 1 \quad \text{if} \quad x\notin A\subset\Lambda;$$

$$G(A,B) = 0 \quad \text{for all other pairs } A,B \text{ with } A\neq B;$$

$$G(A,A) = -\sum_{B\neq A} G(A,B).$$

Then G is the irreducible generator of a semi-group $\{P_t\}_{t\geqslant 0}$. It is easy to check that

$$\sum_{A\subset\Lambda}\mu(A)\,G(A,B) = 0 \quad \text{for all} \quad B\subset\Lambda,$$

and thus μ is the unique equilibrium state of $\{P_t\}_{t\geqslant 0}$. Now let μ_1,μ_2 be two positive elements of $\mathscr{S}(\Lambda)$ and let d_1, G_1 (resp. d_2, G_2) be defined in terms of μ_1 (resp. μ_2) in exactly the same way as d and G were defined in terms of μ.

LEMMA 3.1 Let μ_1, d_1, G_1, μ_2, d_2, G_2 be as above. Suppose there exists a generator

$$\Omega:(\mathscr{P}(\Lambda)\times\mathscr{P}(\Lambda))\times(\mathscr{P}(\Lambda)\times\mathscr{P}(\Lambda))\to\mathbb{R} \quad \text{satisfying:}$$

(i) $\sum_{B_2\subset\Lambda}\Omega(A_1,A_2;\,B_1,B_2) = G_1(A_1,B_1)$ for all $A_1,A_2,B_1\subset\Lambda$.

(ii) $\sum_{B_1\subset\Lambda}\Omega(A_1,A_2;\,B_1,B_2) = G_2(A_2,B_2)$ for all $A_1,A_2,B_2\subset\Lambda$.

(iii) If $A_1\supset A_2$ then $\Omega(A_1,A_2;\,B_1,B_2) = 0$ unless $B_1\supset B_2$.

Then there exists a probability measure ν on $\mathscr{P}(\Lambda)\times\mathscr{P}(\Lambda)$ satisfying the conclusions of Theorem 3.1.

Proof It is a simple matter to check (by induction on n) that (i), (ii) and (iii) imply that for all $n \geqslant 1$ we have

(iv) $\sum_{B_2\subset\Lambda} \Omega^n(A_1, A_2; B_1, B_2) = G_1^n(A_1, B_1)$ for all $A_1, A_2, B_1 \subset \Lambda$.

(v) $\sum_{B_1\subset\Lambda} \Omega^n(A_1, A_2; B_1, B_2) = G_2^n(A_2, B_2)$ for all $A_1, A_2, B_2 \subset \Lambda$.

(vi) If $A_1 \supset A_2$ then $\Omega^n(A_1, B_2; B_1, B_2) = 0$ unless $B_1 \supset B_2$.

(Here of course we define Ω^n, G_1^n and G_2^n by considering them as matrices.) Let $\{P_t\}_{t\geqslant 0}$ be the semi-group on $\mathscr{P}(\Lambda)\times\mathscr{P}(\Lambda)$ that has generator Ω, i.e. $P_t = \exp(t\Omega)$. Define $\nu: \mathscr{P}(\Lambda)\times\mathscr{P}(\Lambda)\to\mathbb{R}$ by

$$\nu(A, B) = \lim_{t\to\infty}\sum_{n=0}^{\infty}\frac{t^n}{n!}\Omega^n(\Lambda, \varnothing; A, B).$$

(By the ergodic theorem for finite Markov chains this limit does exist.) Now as $\Lambda \supset \varnothing$ we have from (vi) that $\nu(A, B) = 0$ unless $A \supset B$. Also we have

$$\sum_{B\subset\Lambda}\nu(A, B) = \lim_{t\to\infty}\sum_{n=0}^{\infty}\frac{t^n}{n!}\sum_{B\subset\Lambda}\Omega^n(\Lambda, \varnothing; A, B)$$
$$= \lim_{t\to\infty}\sum_{n=0}^{\infty}\frac{t^n}{n!}G_1^n(\Lambda, A).$$

Therefore if we let

$$\tilde{\mu}_1(A) = \sum_{B\subset\Lambda}\nu(A, B) \quad \text{then} \quad \tilde{\mu}_1 \in \mathscr{S}(\Lambda),$$

and by the ergodic theorem $\tilde{\mu}_1$ is an equilibrium state of the semi-group with generator G_1. But G_1 is irreducible and so the equilibrium state is unique, thus $\tilde{\mu}_1 = \mu_1$. Similarly we must have

$$\mu_2(B) = \sum_{B\subset\Lambda}\nu(A, B) \quad \text{for all} \quad B \subset \Lambda,$$

and therefore ν has the required properties. □

At this point the reader will have correctly guessed that we will complete the proof of Theorem 3.1 by showing that if μ_1 and μ_2 satisfy the hypotheses of Theorem 3.1 then a generator Ω can be constructed satisfying the hypotheses of Lemma 3.1. However, before we do this we will consider what the generator Ω in

Lemma 3.1 represents. Let Λ_1 and Λ_2 be disjoint copies of Λ, then we can identify $\mathscr{P}(\Lambda) \times \mathscr{P}(\Lambda)$ with $\mathscr{P}(\Lambda_1 \cup \Lambda_2)$ in the obvious way and thus consider Ω as the generator of the semi-group $\{P_t\}_{t \geqslant 0}$ on $\mathscr{P}(\Lambda_1 \cup \Lambda_2)$. If we now observe the Markov chain which has semi-group $\{P_t\}_{t \geqslant 0}$ then it has the properties:

(i) If we just look at what's happening on Λ_1 then we have a Markov chain on $\mathscr{P}(\Lambda_1)$ that is given by the semi-group which has generator G_1.

(ii) Similarly on Λ_2 we have a Markov chain that is given by the semi-group which has generator G_2.

(iii) If we look at both Λ_1 and Λ_2 and if at time t the configuration on Λ_1 is A_1 and the configuration on Λ_2 is A_2 and $A_1 \supset A_2$ then at any time after t the configuration on Λ_2 is also a subset of the configuration on Λ_1, i.e. if $s \geqslant t$ and at time s the configuration on Λ_1 is B_1 and the configuration on Λ_2 is B_2 then $B_1 \supset B_2$.

Thus Ω gives a Markov chain which represents a very particular coupling together of the Markov chains given by G_1 and G_2.

We will now complete the proof of Theorem 3.1 in the manner already stated; thus let $\mu_1, \mu_2 \in \mathscr{S}(\Lambda)$ satisfy the hypotheses of Theorem 3.1. The definition of Ω is given below; it is straightforward to check that Ω does satisfy the hypotheses of Lemma 3.1, and we will leave it to the reader to do so. Let

$$\Omega: (\mathscr{P}(\Lambda) \times \mathscr{P}(\Lambda)) \times (\mathscr{P}(\Lambda) \times \mathscr{P}(\Lambda)) \to \mathbb{R}$$

be given by

$$\Omega(A_1, A_2; A_1 - x, A_2 - x) = \min\,[d_1(x, A_1), d_2(x, A_2)] \text{ if } x \in A_1 \cap A_2,$$

$$\Omega(A_1, A_2; A_1 \cup x, A_2 \cup x) = 1 \quad \text{if } x \in (\Lambda - A_1) \cap (\Lambda - A_2),$$

$$\Omega(A_1, A_2; A_1 - x, A_2) = d_1(x, A_1) - \min\,[d_1(x, A_1), d_2(x, A_2)] \quad \text{if } x \in A_1 \cap A_2,$$

$$\Omega(A_1, A_2; A_1, A_2 - x) = d_2(x, A_2) - \min\,[d_1(x, A_1), d_2(x, A_2)] \quad \text{if } x \in A_1 \cap A_2,$$

$$\Omega(A_1, A_2; A_1 - x, A_2) = d_1(x, A_1) \quad \text{if } x \in A_1, x \notin A_2,$$

$$\Omega(A_1, A_2; A_1 \cup x, A_2) = 1 \quad \text{if } x \notin A_1, x \in A_2,$$

$$\Omega(A_1, A_2; A_1, A_2 - x) = d_2(x, A_2) \quad \text{if } x \in A_2, x \notin A_1,$$

$$\Omega(A_1, A_2; A_1, A_2 \cup x) = 1 \quad \text{if } x \notin A_2, x \in A_1,$$

$\Omega(A_1, A_2; B_1, B_2) = 0$ for all other A_1, A_2, B_1, B_2 with either $A_1 \neq B_1$ or $A_2 \neq B_2$. Finally we define $\Omega(A_1, A_2; A_1, A_2)$ so that $\sum_{B_1 \subset \Lambda} \sum_{B_2 \subset \Lambda} \Omega(A_1, A_2; B_1, B_2) = 0.$ □

NOTES The proof of Theorem 3.1 given here is the same as in Holley (1973c). The original proof of the FKG inequality is given in Fortuin, Kastelyn and Ginibre (1971). For a proof of the ergodic theorem see for example Doob (1953) or Billingsley (1965).

4. *Gibbs states and Markov random fields on countable graphs*

In this chapter we will generalize the states considered in Chapter 1 to the case where the graph has a countably infinite number of vertices, and then prove the analogue of Theorem 1.1. It turns out that the interesting properties of the Gibbs states that we will define do not depend on the structure of the graph. We will thus not prove very much about them here, and the chapter can be considered as an introduction to the next chapters where we examine Gibbs states on countable sets.

Let $\mathscr{G} = (S, e)$ be a graph with vertices S and edges e, with S a countable set, and again let $\mathscr{P}(S)$ denote the set of subsets of S. As in the previous chapters the object of study is certain classes of probability measures on $\mathscr{P}(S)$, but now as $\mathscr{P}(S)$ is an uncountable set we have to give it some more structure in order to define the measures. We will consider $\mathscr{P}(S)$ as a compact Hausdorff space, being the product of $|S|$ copies of the discrete space $\{0, 1\}$. Let $\mathscr{C}(S)$ denote the set of finite subsets of S; if $A \subset B \subset S$ then let

$$[A, B] = \{E \in \mathscr{P}(S) : E \cap B = A\};$$

if $B \in \mathscr{C}(S)$ then we will call $[A, B]$ a *finite dimensional cylinder*. The finite dimensional cylinders from a basis of open sets for the topology on $\mathscr{P}(S)$. Let $\mathscr{F}(S)$ denote the Borel subsets of $\mathscr{P}(S)$, thus $\mathscr{F}(S)$ is the σ-algebra generated by the finite dimensional cylinders. The non-negative measures on $(\mathscr{P}(S), \mathscr{F}(S))$ will be denoted by $\mathscr{M}(S)$, and $\mathscr{S}(S)$ will denote the states on S, i.e. the probability measures on $(\mathscr{P}(S), \mathscr{F}(S))$.

We will define $c: S \times S \to \{0, 1\}$ and ∂A (for any $A \in \mathscr{P}(S)$) just as in the case when the graph was finite. We will assume from now on that ∂x is finite for all $x \in S$; this implies that $\partial A \in \mathscr{C}(S)$

for all $A \in \mathscr{C}(S)$. A function $V: \mathscr{C}(S) \to \mathbb{R}$ will again be called a *potential* if $V(\varnothing) = 0$. We define $J_V: \mathscr{C}(S) \to \mathbb{R}$ by

$$J_V(A) = \sum_{X \subset A} (-1)^{|A-X|} V(X) \quad \text{for all} \quad A \in \mathscr{C}(S),$$

and exactly as in the finite case we have

$$V(A) = \sum_{B \subset A} J_V(B) \quad \text{for all} \quad A \in \mathscr{C}(S).$$

A nearest neighbour potential is defined as in Chapter 1, and Proposition 1.4 remains true (on restricting X to be in $\mathscr{C}(S)$). Let V be a potential on S, let $\Lambda \in \mathscr{C}(S)$ and $B \subset \partial\Lambda$; we define the *Gibbs state on* Λ *with boundary conditions* B *and potential* V to be the state $\pi_\Lambda^B \in \mathscr{S}(\Lambda)$ given by

$$\pi_\Lambda^B(A) = Z^{-1} \exp V(A \cup B) \quad \text{for all} \quad A \subset \Lambda,$$

where
$$Z = \sum_{E \subset \Lambda} \exp V(E \cup B).$$

(As in Chapter 1 we identify the measure π_Λ^B on the finite set $\mathscr{P}(\Lambda)$ with its density.) Note that of course π_Λ^B depends on V, but since the potential V will usually be fixed, no confusion should arise from the notation.

We are now ready to define Gibbs states and Markov random fields on $\mathscr{G}$. Let V be a nearest neighbour potential on S; a state $\mu \in \mathscr{S}(S)$ will be called a *Gibbs state with potential* V if:

(i) $\mu([A, \Lambda]) > 0$ for all finite dimensional cylinders $[A, \Lambda]$.

(ii) If $\Lambda, \Lambda' \in \mathscr{C}(S)$ with $\Lambda \cup \partial\Lambda \subset \Lambda'$, and if $A \subset \Lambda$, $B \subset \Lambda' - \Lambda$, then

$$\frac{\mu([A \cup B, \Lambda'])}{\mu([B, \Lambda' - \Lambda])} = \pi_\Lambda^{B \cap \partial\Lambda}(A).$$

Note that since $[A \cup B, \Lambda'] = [A, \Lambda] \cap [B, \Lambda' - \Lambda]$, (ii) says that the conditional probability that the configuration on Λ is A, given that the configuration on $\Lambda' - \Lambda$ is B, is equal to $\pi_\Lambda^{B \cap \partial\Lambda}(A)$.

A state $\mu \in \mathscr{S}(S)$ will be called a *Markov random field* if:

(i) $\mu([A, \Lambda]) > 0$ for all finite dimensional cylinders $[A, \Lambda]$.

(ii) If $\Lambda, \Lambda' \in \mathscr{C}(S)$ with $\Lambda \cup \partial\Lambda \subset \Lambda'$, and if $A \subset \Lambda$, $B \subset \Lambda' - \Lambda$, then

$$\frac{\mu([A \cup B, \Lambda'])}{\mu([B, \Lambda' - \Lambda])} = \frac{\mu([A \cup (B \cap \partial\Lambda), \Lambda'])}{\mu([B \cap \partial\Lambda, \Lambda' - \Lambda])};$$

i.e. the conditional probability that the configuration on Λ is A, given that the configuration on $\Lambda' - \Lambda$ is B, is the same as the conditional probability that the configuration on Λ is A given that the configuration on $\Lambda' - \Lambda$ is $B \cap \partial\Lambda$.

We leave it to the reader to verify that for finite graphs the above definitions turn out the same as in Chapter 1; at the moment, however, it is not at all clear that Gibbs states and Markov random fields actually exist when the graph is not finite.

We will ignore the question of existence at present and instead prove the analogue of Theorem 1.1.

THEOREM 4.1 Let $\mu \in \mathscr{S}(S)$ with $\mu([A, \Lambda]) > 0$ for all finite dimensional cylinders $[A, \Lambda]$. The following are equivalent:

(i) μ is a Markov random field.

(ii) If $\Lambda \in \mathscr{C}(S)$, $x \notin A \subset \Lambda$ and $x \cup \partial x \subset \Lambda$ then

$$\frac{\mu([A \cup x, \Lambda])}{\mu([A, \Lambda])} = \frac{\mu([(A \cap \partial x) \cup x, \Lambda])}{\mu([(A \cap \partial x), \Lambda])}.$$

(iii) There exists a nearest neighbour potential $V: \mathscr{C}(S) \to \mathbb{R}$ such that μ is a Gibbs state with potential V. Furthermore the potential V in (iii) is uniquely determined by μ.

Proof It is clear that (iii) $\Rightarrow$ (i) $\Rightarrow$ (ii); thus we need only show that (ii) $\Rightarrow$ (iii).

The following lemma will be useful:

LEMMA 4.1 Suppose that (ii) of Theorem 4.1 holds.

Let $\Lambda, \Lambda' \in \mathscr{C}(S)$ with $x \notin A \subset \Lambda$, $x \cup \partial x \subset \Lambda$, and $\Lambda \subset \Lambda'$.

Then
$$\frac{\mu([A \cup x, \Lambda])}{\mu([A, \Lambda])} = \frac{\mu([A \cup x, \Lambda'])}{\mu([A, \Lambda'])}.$$

Assume for a moment the truth of the lemma. From the lemma it easily follows that

if $B \in \mathscr{C}(S)$ and $\Lambda, \Lambda' \in \mathscr{C}(S)$ with $B \cup \partial B \subset \Lambda$, $B \cup \partial B \subset \Lambda'$,

then
$$\frac{\mu([B, \Lambda])}{\mu([\varnothing, \Lambda])} = \frac{\mu([B, \Lambda'])}{\mu([\varnothing, \Lambda'])}.$$

We can thus define $V: \mathscr{C}(S) \to \mathbb{R}$ by putting

$$V(B) = \log\left[\frac{\mu([B, \Lambda])}{\mu([\varnothing, \Lambda])}\right] \quad \text{for any } \Lambda \in \mathscr{C}(S) \text{ with } \Lambda \supset B \cup \partial B.$$

It follows fairly easily from Proposition 1.4 that V is a nearest neighbour potential. (We leave this for the reader to check.) Now let $\Lambda, \Lambda' \in \mathscr{C}(S)$ with $\Lambda \cup \partial\Lambda \subset \Lambda'$, let $A \subset \Lambda$ and $B \subset \Lambda' - \Lambda$. Take any $\Lambda'' \in \mathscr{C}(S)$ with $\Lambda' \cup \partial\Lambda' \subset \Lambda''$. Then if $E \subset \Lambda$ we have by Lemma 4.1 that

$$\begin{aligned}\frac{\mu([E \cup B, \Lambda'])}{\mu([B, \Lambda'])} &= \frac{\mu([E \cup B, \Lambda''])}{\mu([B, \Lambda''])} \\ &= \frac{\mu([E \cup B, \Lambda''])}{\mu([\varnothing, \Lambda''])} \frac{\mu([\varnothing, \Lambda''])}{\mu([B, \Lambda''])} \\ &= \exp[V(E \cup B) - V(B)].\end{aligned}$$

Therefore

$$\begin{aligned}\frac{\mu([A \cup B, \Lambda'])}{\mu([B, \Lambda' - \Lambda])} &= \frac{\mu([A \cup B, \Lambda'])}{\mu([B, \Lambda'])} \frac{\mu([B, \Lambda'])}{\sum\limits_{E \subset \Lambda} \mu([E \cup B, \Lambda'])} \\ &= \frac{\exp[V(A \cup B) - V(B)]}{\sum\limits_{E \subset \Lambda} \exp[V(E \cup B) - V(B)]} = \pi_\Lambda^B(A) = \pi_\Lambda^{B \cap \partial\Lambda}(A).\end{aligned}$$

Thus μ is a Gibbs state with potential V, and also it is clear that V is uniquely determined by μ. Finally we must give a proof of Lemma 4.1: Let $\Lambda, \Lambda' \in \mathscr{C}(S)$, $x \notin A \subset \Lambda$ with $x \cup \partial x \subset \Lambda$ and $\Lambda \subset \Lambda'$. Then

$$\begin{aligned}\mu([A \cup x, \Lambda]) &= \sum_{E \subset \Lambda' - \Lambda} \mu([E \cup A \cup x, \Lambda']) \\ &= \sum_{E \subset \Lambda' - \Lambda} \frac{\mu([E \cup A \cup x, \Lambda'])}{\mu([E \cup A, \Lambda'])} \mu([E \cup A, \Lambda']) \\ &= \sum_{E \subset \Lambda' - \Lambda} \frac{\mu([A \cup x, \Lambda'])}{\mu([A, \Lambda')]} \mu([E \cup A, \Lambda']) \\ &= \frac{\mu([A \cup x, \Lambda'])}{\mu([A, \Lambda'])} \mu([A, \Lambda]),\end{aligned}$$

and thus $$\frac{\mu([A \cup x, \Lambda])}{\mu([A, \Lambda])} = \frac{\mu([A \cup x, \Lambda'])}{\mu([A, \Lambda'])}. \square$$

The above theorem would of course be of no value if Gibbs states did not exist; however if V is a nearest neighbour potential on S then there does exist a Gibbs state with potential V. We leave the proof of this fact to Chapter 5 (in order to avoid repeating almost the same proof twice).

In the case of a finite graph there is exactly one Gibbs state with potential V for a given potential V; but when the graph is not finite we may not have this uniqueness, i.e. for some potentials V there may be more than one Gibbs state with potential V. When this happens we say that *phase transition* occurs for V. A lot of the rest of the book is taken up with trying to determine when phase transition does occur.

In the definition of a Gibbs state we assumed that the measure of every finite dimensional cylinder was strictly positive. This was not in fact necessary as it follows automatically from the rest of the definition:

PROPOSITION 4.1 Let V be a nearest neighbour potential on S. If $\mu \in \mathscr{S}(S)$ then the following are equivalent:

(i) μ is a Gibbs state with potential V.

(ii) If $\Lambda, \Lambda' \in \mathscr{C}(S)$ with $\Lambda \cup \partial\Lambda \subset \Lambda'$, if $A \subset \Lambda$ and $B \subset \Lambda' - \Lambda$ then
$$\mu([A \cup B, \Lambda']) = \mu([B, \Lambda' - \Lambda])\, \pi_\Lambda^{B \cap \partial\Lambda}(A).$$

Proof Clearly (i) $\Rightarrow$ (ii) and to show that (ii) $\Rightarrow$ (i) it is only necessary to show that (ii) implies that $\mu([A, \Lambda]) > 0$ for every finite dimensional cylinder $[A, \Lambda]$. Thus suppose that (ii) holds and let $\Lambda \in \mathscr{C}(S)$ and $A \subset \Lambda$. Then if $\Lambda' = \Lambda \cup \partial\Lambda$ we must have

$$\mu([A, \Lambda]) = \sum_{B \subset \partial\Lambda} \mu([A \cup B, \Lambda']) = \sum_{B \subset \partial\Lambda} \mu([B, \partial\Lambda])\, \pi_\Lambda^B(A)$$
$$\geqslant [\min_{B \subset \partial\Lambda} \pi_\Lambda^B(A)] \sum_{B \subset \partial\Lambda} \mu([B, \partial\Lambda]) = \min_{B \subset \partial\Lambda} \pi_\Lambda^B(A) > 0. \square$$

The definition of a Markov random field given in this chapter is really the analogue to the definition of a nearest neighbour state given in Chapter 1. However we could define a Markov random field on S by a definition which is a closer analogue to the definition in Chapter 1 since we have:

PROPOSITION 4.2 Let $\mu \in \mathscr{S}(S)$ with $\mu([A, \Lambda]) > 0$ for all finite dimensional cylinders $[A, \Lambda]$. The following are equivalent:

(i) μ is a Markov random field.

(ii) If $\Lambda, \Lambda' \in \mathscr{C}(S)$ with $\Lambda \cup \partial\Lambda \subset \Lambda'$, and if $A \subset \Lambda, B \subset \Lambda' - \Lambda$ then

$$\frac{\mu([A \cup B, \Lambda'])}{\mu([B, \Lambda' - \Lambda])} = \frac{\mu([A \cup (B \cap \partial\Lambda), \Lambda \cup \partial\Lambda])}{\mu([B \cap \partial\Lambda, \partial\Lambda])},$$

i.e. the conditional probability that the configuration is A on Λ, given that the configuration is B on $\Lambda' - \Lambda$, is equal to the conditional probability that the configuration is A on Λ, given that the configuration is $B \cap \partial\Lambda$ on $\partial\Lambda$.

Proof We leave this to the reader. □

We now give a characterization of Gibbs states that will serve as some motivation for the next chapter. For $A \in \mathscr{P}(S)$ let $r_A: \mathscr{P}(S) \to \mathscr{P}(A)$ be the projection defined by

$$r_A(X) = X \cap A \quad \text{for} \quad X \in \mathscr{P}(S).$$

We let r_A induce a mapping (also denoted by r_A) from $\mathscr{M}(S)$ to $\mathscr{M}(A)$ by defining

$$r_A(\mu)(\Omega) = \mu(r_A^{-1}(\Omega)) \quad \text{for} \quad \Omega \in \mathscr{F}(A), \mu \in \mathscr{M}(S).$$

If $A \in \mathscr{P}(S)$ and $B \subset S - A$ then we define $r_A^B: \mathscr{M}(S) \to \mathscr{M}(A)$ by letting

$$r_A^B(\mu)(\Omega) = \mu(\{X \in \mathscr{P}(S): X \cap A \in \Omega, X \cap (S - A) = B\})$$
$$\text{for all} \quad \Omega \in \mathscr{F}(A), \mu \in \mathscr{M}(S).$$

Now let V be a nearest neighbour potential on S. For $\Lambda \in \mathscr{C}(S)$ define $f^\Lambda: \mathscr{P}(\Lambda) \times \mathscr{P}(S - \Lambda) \to \mathbb{R}$ by

$$f^\Lambda(A, X) = \pi_\Lambda^{X \cap \partial\Lambda}(A).$$

It is not difficult to show that $f^\Lambda \in C(\mathscr{P}(\Lambda) \times \mathscr{P}(S - \Lambda))$ (where for any topological space Y, $C(Y)$ will denote the vector space of continuous functions from Y to $\mathbb{R}$). With the above definitions we can now prove:

PROPOSITION 4.3 μ is a Gibbs state with potential V if and only if for all $\Lambda \in \mathscr{C}(S)$ and $A \subset \Lambda$ we have

$$r_{S-\Lambda}^A(\mu) = f^\Lambda(A, \cdot)\, r_{S-\Lambda}(\mu).$$

Proof Suppose that for all $\Lambda \in \mathscr{C}(S)$ and $A \subset \Lambda$ we do have $r^A_{S-\Lambda}(\mu) = f^\Lambda(A, \cdot)\, r_{S-\Lambda}(\mu)$. Let $\Lambda, \Lambda' \in \mathscr{C}(S)$ with $\Lambda \cup \partial\Lambda \subset \Lambda'$ and let $A \subset \Lambda, B \subset \Lambda' - \Lambda$. Then

$$\begin{aligned} r^A_{S-\Lambda}(\mu)([B, \Lambda' - \Lambda]) &= \int_{[B,\Lambda'-\Lambda]} f^\Lambda(A, X)\, \mathrm{d}r_{S-\Lambda}(\mu)(X) \\ &= f^\Lambda(A, B) \int_{[B,\Lambda'-\Lambda]} \mathrm{d}r_{S-\Lambda}(\mu)(X) \\ &= f^\Lambda(A, B)\, r_{S-\Lambda}(\mu)([B, \Lambda' - \Lambda]), \end{aligned}$$

(since $f^\Lambda(A, X) = f^\Lambda(A, B)$ for all $X \in [B, \Lambda' - \Lambda]$).

But $$r^A_{S-\Lambda}(\mu)([B, \Lambda' - \Lambda]) = \mu([A \cup B, \Lambda'])$$

and $$r_{S-\Lambda}(\mu)([B, \Lambda' - \Lambda]) = \mu([B, \Lambda' - \Lambda]);$$

also $$f^\Lambda(A, B) = \pi^{B \cap \partial\Lambda}_\Lambda(A).$$

Therefore $$\mu([A \cup B, \Lambda']) = \mu([B, \Lambda' - \Lambda])\, \pi^{B \cap \partial\Lambda}_\Lambda(A),$$

and thus by Proposition 4.1 μ is a Gibbs state with potential V. Conversely, if μ is a Gibbs state with potential V then the measures $r^A_{S-\Lambda}(\mu)$ and $f^\Lambda(A, \cdot)\, r_{S-\Lambda}(\mu)$ agree on all finite dimensional cylinders of the form $[B, \Lambda' - \Lambda]$ with $\Lambda' \supset \Lambda \cup \partial\Lambda$ and $B \subset \Lambda' - \Lambda$. Thus $r^A_{S-\Lambda}(\mu)$ and $f^\Lambda(A, \cdot)\, r_{S-\Lambda}(\mu)$ agree on all finite dimensional cylinders of $\mathscr{P}(S - \Lambda)$ and hence are the same. □

We conclude this chapter by noting that the functions $\{f^\Lambda\}_{\Lambda \in \mathscr{C}(S)}$ satisfy:

(i) $f^\Lambda(A, X) \geqslant 0$ for all $A \in \mathscr{P}(\Lambda), X \in \mathscr{P}(S - \Lambda)$.

(ii) $\sum_{A \subset \Lambda} f^\Lambda(A, X) = 1$ for all $X \in \mathscr{P}(S - \Lambda)$.

(iii) If $\Lambda \subset \Lambda' \in \mathscr{C}(S)$, $A \subset \Lambda$, $B \subset \Lambda' - \Lambda$ and $X \subset S - \Lambda'$ then

$$f^{\Lambda'}(A \cup B, X) = f^\Lambda(A, B \cup X) \sum_{C \subset \Lambda} f^{\Lambda'}(C \cup B, X).$$

(Clearly (i) and (ii) hold, and (iii) is easily seen to be true.) In the next chapter we will consider functions $\{f^\Lambda\}_{\Lambda \in \mathscr{C}(S)}$ having these properties.

Notes As mentioned in Chapter 1, Markov random fields on $\mathbb{Z}^\nu$ were introduced by Dobrushin (1968a) and the definition for arbitrary countable graphs is essentially the same. Classically Gibbs states were only considered on finite subsets of $\mathbb{Z}^\nu$; various thermodynamic quantities were computed and their limiting behaviour examined. The idea of defining a Gibbs state as a measure on $\mathscr{P}(\mathbb{Z}^\nu)$ was introduced by Dobrushin, Minlos and Ruelle. (Their definitions all appeared at about the same time.) The definition of a Gibbs state on $\mathbb{Z}^\nu$ due to Dobrushin (1968b, 1968c, 1969) (for translation invariant pair potentials on $\mathbb{Z}^\nu$) is the one used here. Minlos (1967a, 1967b) defined a Gibbs state on $\mathbb{Z}^\nu$ as a state obtained by taking limits of Gibbs states with boundary conditions on finite subsets of $\mathbb{Z}^\nu$. Dobrushin (1968b) shows that his definition is the same as that of Minlos. In Ruelle (1967a, 1967b) a Gibbs state on $\mathbb{Z}^\nu$ is defined as a translation invariant state with maximum entropy amongst all translation invariant states with a given density and energy. Lanford and Ruelle (1969) show that this definition is the same as that given by Proposition 4.3 and is thus equivalent to Dobrushin's definition.

The equivalence of Markov random fields and Gibbs states with nearest neighbour potentials on $\mathbb{Z}^\nu$, under the assumption of translation invariance, was shown by Spitzer (1971b).

A lot of work is being done at the moment on what could be considered a generalization of Chapter 2, namely to construct Markov processes whose equilibrium states are Gibbs states with given potentials. Those readers interested in this and related topics should look at the work of Spitzer (1970), Holley (1970, 1971, 1973a, 1973b), Liggett (1972, 1973), Harris (1972), Dobrushin (1971) and Dobrushin, Patetski-Shapiro and Vasilev (1969).

5. *Gibbs states on countable sets*

In this chapter we will examine certain probability measures on $\mathscr{P}(S)$, the set of all subsets of S, where S is a countable set. We will be interested in the existence and uniqueness of measures satisfying certain properties. As before we will regard the points of S as sites, each of which can be either empty or occupied by a particle (or some other entity); the subset $A \in \mathscr{P}(S)$ will be regarded as describing the situation when the points of A are occupied and the points of $S-A$ are empty. The elements of $\mathscr{P}(S)$ will sometimes be called configurations and the probability measures on $\mathscr{P}(S)$ that we will look at will describe the equilibrium distribution of the configurations of some model.

The basic assumption concerning the models that we will consider is the following: let Λ be a finite subset of S, $A \subset \Lambda$ and $X \subset S-\Lambda$; we will suppose that the model is such that the conditional probability of there being particles on Λ at exactly the points of A, given that on $S-\Lambda$ there are particles at exactly the points of X, is specified. (This says that if we know what is happening outside a finite subset of S then we can compute the distribution of particles inside the finite set.) Let us denote the above conditional probability by $f^{\Lambda}(A, X)$; our aim in this chapter is firstly to determine what relations the $f^{\Lambda}(A, X)$ must satisfy and secondly, to find out whether the $f^{\Lambda}(A, X)$ uniquely determine a probability measure on $\mathscr{P}(S)$.

The main examples of models of the type described above arise in classical statistical mechanics and thus the approach we will take to the material will draw heavily on the terminology and techniques of statistical mechanics.

We require the following notation and definitions, some of which have already appeared in Chapter 4. In an attempt to keep this chapter self-contained we will write them out again (and apologize to those readers who have read Chapter 4). S will be a countable set and $\mathscr{P}(S)$ will denote the set of all subsets of S. We will regard $\mathscr{P}(S)$ as a compact Hausdorff space, being the product of $|S|$ copies of the discrete space $\{0,1\}$. $\mathscr{C}(S)$ will denote the finite subsets of S. If $A \subset B \in \mathscr{P}(S)$ then let

$$[A,B] = \{X \in \mathscr{P}(S): X \cap B = A\}.$$

If $B \in \mathscr{C}(S)$ then $[A,B]$ is called a *finite dimensional cylinder*. The finite dimensional cylinders form a basis of open sets for the topology on $\mathscr{P}(S)$. Let $\mathscr{F}(S)$ denote the Borel subsets of $\mathscr{P}(S)$, thus $\mathscr{F}(S)$ is the σ-algebra generated by the finite dimensional cylinders. We let $\mathscr{M}(S)$ be the set of non-negative measures on $(\mathscr{P}(S), \mathscr{F}(S))$ considered as a topological space with the vague topology (i.e. the relative topology obtained by regarding $\mathscr{M}(S)$ as a subset of the dual of $C(\mathscr{P}(S))$). $\mathscr{S}(S)$ will denote the set of states on S (i.e. the set of probability measures on $(\mathscr{P}(S), \mathscr{F}(S))$). Elements of $\mathscr{S}(S)$ are often also called *random fields*. Recall that $\mathscr{S}(S)$ is a compact subset of $\mathscr{M}(S)$ and that if $\mu_n, \mu \in \mathscr{S}(S)$ then $\mu_n \to \mu$ in the vague topology as $n \to \infty$ if and only if

$$\mu_n([A,\Lambda]) \to \mu([A,\Lambda])$$

as $n \to \infty$ for every finite dimensional cylinder $[A,\Lambda]$.

For $A \in \mathscr{P}(S)$ let $r_A: \mathscr{P}(S) \to \mathscr{P}(A)$ be the projection defined by

$$r_A(X) = X \cap A \quad \text{for all} \quad X \in \mathscr{P}(S).$$

It is clear that r_A is continuous. We let r_A induce a mapping (also denoted by r_A) from $\mathscr{M}(S)$ to $\mathscr{M}(A)$ defined by

$$r_A(\mu)(\Omega) = \mu(r_A^{-1}(\Omega)) \quad \text{for all} \quad \Omega \in \mathscr{F}(A), \mu \in \mathscr{M}(S).$$

Note that r_A maps $\mathscr{S}(S)$ into $\mathscr{S}(A)$. If $A \in \mathscr{P}(S)$ and $B \subset S - A$ then we define $r_A^B: \mathscr{M}(S) \to \mathscr{M}(A)$ by

$$r_A^B(\mu)(\Omega) = \mu(\{X \in \mathscr{P}(S): X \cap A \in \Omega, X \cap (S-A) = B\})$$
$$\text{for all} \quad \Omega \in \mathscr{F}(A), \mu \in \mathscr{M}(S).$$

If $\Lambda \in \mathscr{C}(S)$ then we have

$$r_{S-\Lambda}(\mu) = \sum_{A \subset \Lambda} r_{S-\Lambda}^A(\mu).$$

Finally, if Y is a topological space then $C(Y)$ will denote the vector space of continuous functions from Y to $\mathbb{R}$.

We are now in a position to define a Gibbs state on S. For each $\Lambda \in \mathscr{C}(S)$ let $f^{\Lambda} \in C(\mathscr{P}(\Lambda) \times \mathscr{P}(S-\Lambda))$; we will say that $\{f^{\Lambda}\}_{\Lambda \in \mathscr{C}(S)}$ is a *local specification* if:

(i) $f^{\Lambda}(A, X) \geqslant 0$ for all $A \in \mathscr{P}(\Lambda), X \in \mathscr{P}(S-\Lambda)$.

(ii) $\sum_{A \subset \Lambda} f^{\Lambda}(A, X) = 1$ for all $X \in \mathscr{P}(S-\Lambda)$.

(iii) If $\Lambda \subset \Lambda' \in \mathscr{C}(S)$, $A \subset \Lambda$, $B \subset \Lambda' - \Lambda$, and $X \subset S - \Lambda'$ then

$$f^{\Lambda'}(A \cup B, X) = f^{\Lambda}(A, B \cup X) \sum_{C \subset \Lambda} f^{\Lambda'}(C \cup B, X).$$

Let $\mathscr{V} = \{f^{\Lambda}\}_{\Lambda \in \mathscr{C}(S)}$ be a local specification; we will call $\mu \in \mathscr{S}(S)$ a *Gibbs state with specification* $\mathscr{V}$ if for all $\Lambda \in \mathscr{C}(S)$, $A \subset \Lambda$ we have

$$r^{A}_{S-\Lambda}(\mu) = f^{\Lambda}(A, \cdot\,)\, r_{S-\Lambda}(\mu).$$

Let $\mathscr{G}(\mathscr{V}) = \{\mu \in \mathscr{S}(S)$: μ is a Gibbs state with specification $\mathscr{V}\}$. Thus a random field μ is in $\mathscr{G}(\mathscr{V})$ if for all $\Lambda \in \mathscr{C}(S)$ and $A \subset \Lambda$ we have that the Radon–Nikodym derivative of $r^{A}_{S-\Lambda}(\mu)$ with respect to $r_{S-\Lambda}(\mu)$ is the specified function $f^{\Lambda}(A, \cdot\,)$.

The functions f^{Λ} given by a local specification are in fact the same as the conditional probabilities mentioned at the beginning of this chapter. To see this we need the following definitions:

If $A \in \mathscr{P}(S)$ then by our previous notation $\mathscr{F}(A)$ denotes the Borel subsets of the compact Hausdorff space $\mathscr{P}(A)$; we let $\overline{\mathscr{F}}(A)$ denote the sub σ-algebra of $\mathscr{F}(S)$ generated by

$$\{[B, \Lambda]: B \subset \Lambda \in \mathscr{C}(A)\}.$$

$\overline{\mathscr{F}}(A)$ consists of those elements of $\mathscr{F}(S)$ that are of the form

$$\{X \cup B: X \in \Omega, B \subset S - A\} \quad \text{for some} \quad \Omega \in \mathscr{F}(A).$$

It is easily seen that a function $g: \mathscr{P}(S) \to \mathbb{R}$ is $\overline{\mathscr{F}}(A)$-measurable if and only if there exists an $\mathscr{F}(A)$-measurable $g': \mathscr{P}(A) \to \mathbb{R}$ such that

$$g(X) = g'(X \cap A) \quad \text{for all} \quad X \in \mathscr{P}(S).$$

For $\mu \in \mathscr{S}(S)$, $\Omega \in \mathscr{F}(S)$ and $B \in \mathscr{P}(S)$ let $P_{\mu}[\Omega | \overline{\mathscr{F}}(B)]$ denote the conditional probability (with respect to μ) of Ω given $\overline{\mathscr{F}}(A)$; i.e.

$P_\mu[\Omega \mid \overline{\mathscr{F}}(B)]: \mathscr{P}(S) \to \mathbb{R}$ is $\overline{\mathscr{F}}(B)$-measurable and for all $\Omega' \in \overline{\mathscr{F}}(B)$ we have

$$\int_{\Omega'} P_\mu[\Omega \mid \overline{\mathscr{F}}(B)]\,\mathrm{d}\mu = \mu(\Omega' \cap \Omega).$$

Now let $\Lambda \in \mathscr{C}(S)$, $A \subset \Lambda$ and let $g^\Lambda(A, \cdot): \mathscr{P}(S-\Lambda) \to \mathbb{R}$ be the $\mathscr{F}(S-\Lambda)$-measurable function such that

$$P_\mu[[A,\Lambda] \mid \overline{\mathscr{F}}(S-\Lambda)](X) = g^\Lambda(A, (S-\Lambda) \cap X) \text{ for all } X \in \mathscr{P}(S).$$

(Since $P_\mu[[A,\Lambda] \mid \overline{\mathscr{F}}(S-\Lambda)]$ is determined μ-a.e. it follows that $g^\Lambda(A, \cdot)$ is determined $r_{S-\Lambda}(\mu)$-a.e.) Then if $B \subset \Lambda' \in \mathscr{C}(S-\Lambda)$ we have

$$r^A_{S-\Lambda}(\mu)([B,\Lambda']) = \mu([A,\Lambda] \cap [B,\Lambda'])$$

$$= \int_{[B,\Lambda']} P_\mu[[A,\Lambda] \mid \overline{\mathscr{F}}(S-\Lambda)]\,\mathrm{d}\mu = \int_{[B,\Lambda']} g^\Lambda(A,X)\,\mathrm{d}r_{S-\Lambda}(\mu)(X).$$

(Note that we have used $[B,\Lambda']$ to denote both an element of $\mathscr{F}(S)$ and an element of $\mathscr{F}(S-\Lambda)$.) Therefore

$$r^A_{S-\Lambda}(\mu) = g^\Lambda(A, \cdot)\, r_{S-\Lambda}(\mu);$$

hence a local specification can be regarded as specifying what $P_\mu[[A,\Lambda] \mid \overline{\mathscr{F}}(S-\Lambda)]$ has to be for all $A \subset \Lambda \in \mathscr{C}(S)$.

The conditions (i), (ii) and (iii) that define a local specification arise in a natural way, since we have:

PROPOSITION 5.1 Let $\mu \in \mathscr{S}(S)$ with $\mu([A,\Lambda]) > 0$ for all finite dimensional cylinders $[A,\Lambda]$. For $A \subset \Lambda \in \mathscr{C}(S)$ let $g^\Lambda(A, \cdot)$ be the Radon–Nikodym derivative of $r^A_{S-\Lambda}(\mu)$ with respect to $r_{S-\Lambda}(\mu)$. Suppose we can take g^Λ to be continuous (i.e.

$$g^\Lambda \in C(\mathscr{P}(\Lambda) \times \mathscr{P}(S-\Lambda))).$$

Then $\{g^\Lambda\}_{\Lambda \in \mathscr{C}(S)}$ is a local specification.

Proof Let $\Lambda \in \mathscr{C}(S)$ and let $u_1, u_2 \in C(\mathscr{P}(S-\Lambda))$. Then since $\mu([A,B]) > 0$ for every finite dimensional cylinder $[A,B]$ we have that $u_1 = u_2$ $r_{S-\Lambda}(\mu)$-a.e. if and only if $u_1(X) = u_2(X)$ for all $X \in \mathscr{P}(S-\Lambda)$.

Thus we immediately have that

$$g^{\Lambda}(A, X) \geqslant 0 \quad \text{for all} \quad A \subset \Lambda, X \in \mathscr{P}(S-\Lambda),$$

and
$$\sum_{A \subset \Lambda} g^{\Lambda}(A, X) = 1 \quad \text{for all} \quad X \in \mathscr{P}(S-\Lambda).$$

Now let $\Lambda \subset \Lambda' \in \mathscr{C}(S)$, $A \subset \Lambda, B \subset \Lambda' - \Lambda$ and $E \subset F \subset \mathscr{C}(S-\Lambda')$. Then

$$\int_{[E,F]} g^{\Lambda'}(A \cup B, X)\, \mathrm{d}r_{S-\Lambda'}(\mu)(X) = \int_{[E,F]} \mathrm{d}r^{A \cup B}_{S-\Lambda'}(\mu)(X)$$

$$= \mu([A \cup B \cup E, \Lambda' \cup F]) = \int_{[E \cup B, F \cup (\Lambda'-\Lambda)]} \mathrm{d}r^{A}_{S-\Lambda}(\mu)(Y)$$

$$= \int_{[E \cup B, F \cup (\Lambda'-\Lambda)]} g^{\Lambda}(A, Y)\, \mathrm{d}r_{S-\Lambda}(\mu)(Y)$$

$$= \int_{[E \cup B, F \cup (\Lambda'-\Lambda)]} g^{\Lambda}(A, Y) \sum_{C \subset \Lambda} \mathrm{d}r^{C}_{S-\Lambda}(\mu)(Y)$$

$$= \int_{[E,F]} g^{\Lambda}(A, B \cup X) \sum_{C \subset \Lambda} \mathrm{d}r^{B \cup C}_{S-\Lambda'}(\mu)(X)$$

$$= \int_{[E,F} g^{\Lambda}(A, B \cup X) \sum_{C \subset \Lambda} g^{\Lambda'}(B \cup C, X)\, \mathrm{d}r_{S-\Lambda'}(\mu)(X).$$

Therefore

$$g^{\Lambda'}(A \cup B, X) = g^{\Lambda}(A, B \cup X) \sum_{C \subset \Lambda} g^{\Lambda'}(B \cup C, X)$$

for $r_{S-\Lambda'}(\mu)$-a.e. $X \in \mathscr{P}(S-\Lambda')$ and thus

$$g^{\Lambda'}(A \cup B, X) = g^{\Lambda}(A, B \cup X) \sum_{C \subset \Lambda} g^{\Lambda'}(B \cup C, X) \quad \text{for all} \quad X \in \mathscr{P}(S-\Lambda').$$

Hence $\{g^{\Lambda}\}_{\Lambda \in \mathscr{C}(S)}$ is a local specification. □

We now show that Gibbs states with a given specification always exist (and thus in particular clear up the existence problem left over from Chapter 2).

Proposition 5.2 Let $\mathscr{V} = \{f^{\Lambda}\}_{\Lambda \in \mathscr{C}(S)}$ be a local specification. Then $\mathscr{G}(\mathscr{V})$ is a non-empty, convex compact subset of $\mathscr{S}(S)$.

Proof It is clear that $\mathscr{G}(\mathscr{V})$ is convex and compact (since in this case being compact is the same as being closed) so we need only show that $\mathscr{G}(\mathscr{V})$ is non-empty. Let $\Lambda_n \in \mathscr{C}(S)$ with $\Lambda_n \uparrow S$ and let $\mu_n \in \mathscr{S}(S)$ be defined by

$$\mu_n([A, \Lambda_n]) = f^{\Lambda_n}(A, \varnothing) \quad \text{for all} \quad A \subset \Lambda_n,$$
$$\mu_n([\varnothing, x]) = 1 \quad \text{for all} \quad x \notin \Lambda_n.$$

By choosing a subsequence we can assume that there exists $\mu \in \mathscr{S}(S)$ such that $\mu_n \to \mu$ (in the vague topology) as $n \to \infty$. Now let $A \subset \Lambda \in \mathscr{C}(S)$ and $E \subset F \in \mathscr{C}(S - \Lambda)$. Then if n is such that $\Lambda_n \supset \Lambda \cup F$ we have

$$r^A_{S-\Lambda}(\mu_n)([E, F]) = \sum_{B \subset \Lambda_n - (\Lambda \cup F)} \mu_n([A \cup E \cup B, \Lambda_n])$$
$$= \sum_{B \subset \Lambda_n - (\Lambda \cup F)} f^{\Lambda_n}(A \cup E \cup B, \varnothing).$$

We also have

$$\int_{[E, F]} f^{\Lambda}(A, X)\, \mathrm{d}r_{S-\Lambda}(\mu_n)(X) = \sum_{C \subset \Lambda} \int_{[E, F]} f^{\Lambda}(A, X)\, \mathrm{d}r^C_{S-\Lambda}(\mu_n)(X)$$
$$= \sum_{C \subset \Lambda} \sum_{B \subset \Lambda_n - (\Lambda \cup F)} f^{\Lambda}(A, E \cup B)\, \mu_n([C \cup E \cup B, \Lambda_n])$$
$$= \sum_{C \subset \Lambda} \sum_{B \subset \Lambda_n - (\Lambda \cup F)} f^{\Lambda}(A, E \cup B) f^{\Lambda_n}(C \cup E \cup B, \varnothing).$$

But $\quad \sum_{C \subset} f^{\Lambda}(A, E \cup B) f^{\Lambda_n}(C \cup E \cup B, \varnothing) = f^{\Lambda_n}(A \cup E \cup B, \varnothing)$

and thus

$$\int_{[E, F]} f^{\Lambda}(A, X)\, \mathrm{d}r_{S-\Lambda}(\mu_n)(X) = \sum_{B \subset \Lambda_n - (\Lambda \cup F)} f^{\Lambda_n}(A \cup E \cup B, \varnothing)$$
$$= r^A_{S-\Lambda}(\mu_n)([E, F]).$$

Letting $n \to \infty$ we get

$$r^A_{S-\Lambda}(\mu)([E, F]) = \int_{[E, F]} f^{\Lambda}(A, X)\, \mathrm{d}r_{S-\Lambda}(\mu)(X)$$

and hence $r^A_{S-\Lambda}(\mu) = f^{\Lambda}(A, \cdot)\, r_{S-\Lambda}(\mu)$. Therefore $\mu \in \mathscr{G}(\mathscr{V})$, i.e. $\mathscr{G}(\mathscr{V})$ is non-empty. □

We will say that a local specification $\mathscr{V} = \{f^\Lambda\}_{\Lambda \in \mathscr{C}(S)}$ is *positive* if $f^\Lambda(A, X) > 0$ for all $A \subset \Lambda \in \mathscr{C}(S)$ and $X \in \mathscr{P}(S - \Lambda)$. Note that since $\mathscr{P}(S - \Lambda)$ and $\mathscr{P}(\Lambda)$ are compact, then $\mathscr{V}$ positive implies that for each $\Lambda \in \mathscr{C}(S)$ there exists $\delta_\Lambda > 0$ such that $f^\Lambda(A, X) \geqslant \delta_\Lambda$ for all $A \subset \Lambda$ and $X \in \mathscr{P}(S - \Lambda)$. The functions f^Λ that occur in statistical mechanics are usually in the form of the exponential of a potential function. We will now show that any positive local specification can be put in this form. A function

$$V: \mathscr{C}(S) \to \mathbb{R}$$

will be called a *potential* on S if $V(\varnothing) = 0$.

PROPOSITION 5.3 Let $\mathscr{V} = \{f^\Lambda\}_{\Lambda \in \mathscr{C}(S)}$ be a positive local specification. Then there exists a unique potential V on S such that if $\Lambda \in \mathscr{C}(S)$, $A \subset \Lambda$ and $B \subset \mathscr{C}(S - \Lambda)$ then

$$f^\Lambda(A, B) = Z_{\Lambda, B}^{-1} \exp V(A \cup B),$$

where

$$Z_{\Lambda, B} = \sum_{E \subset \Lambda} \exp V(E \cup B).$$

Proof It is easy to see that if V exists then it is unique. If $\Lambda \subset \Lambda' \in \mathscr{C}(S)$ and $A \subset \Lambda$ then

$$f^{\Lambda'}(A, \varnothing) = f^\Lambda(A, \varnothing) \sum_{C \subset \Lambda} f^{\Lambda'}(C, \varnothing)$$

and thus

$$\frac{f^{\Lambda'}(A, \varnothing)}{f^{\Lambda'}(\varnothing, \varnothing)} = \frac{f^\Lambda(A, \varnothing)}{f^\Lambda(\varnothing, \varnothing)}.$$

Hence we can define a function $W: \mathscr{C}(S) \to \mathbb{R}$ by letting

$$W(A) = \frac{f^\Lambda(A, \varnothing)}{f^\Lambda(\varnothing, \varnothing)}$$

for any $\Lambda \in \mathscr{C}(S)$ with $\Lambda \supset A$. Since $W(A) > 0$ we can define $V: \mathscr{C}(S) \to \mathbb{R}$ by putting

$$V(A) = \log W(A) \quad \text{for} \quad A \in \mathscr{C}(S).$$

V is a potential on S since $W(\varnothing) = 1$. Now for

$$\Lambda \in \mathscr{C}(S), \quad B \in \mathscr{C}(S - \Lambda)$$

let

$$Z_{\Lambda, B} = \sum_{A \subset \Lambda} W(A \cup B).$$

Let $\Lambda \in \mathscr{C}(S)$, $A \subset \Lambda$, $B \in \mathscr{C}(S-\Lambda)$ and choose $\Lambda' \in \mathscr{C}(S)$ with $\Lambda \cup B \subset \Lambda'$. Then we have

$$\frac{f^{\Lambda}(A,B)\,Z_{\Lambda,B}}{W(A \cup B)} = \frac{f^{\Lambda}(A,B)}{W(A \cup B)} \sum_{E \subset \Lambda} W(E \cup B)$$

$$= \frac{f^{\Lambda}(A,B)}{f^{\Lambda'}(A \cup B, \varnothing)} \sum_{E \subset \Lambda} f^{\Lambda'}(E \cup B, \varnothing) = \frac{f^{\Lambda'}(A \cup B, \varnothing)}{f^{\Lambda'}(A \cup B, \varnothing)} = 1.$$

Therefore

$$f^{\Lambda}(A,B) = Z_{\Lambda,B}^{-1}\, W(A \cup B) = Z_{\Lambda,B}^{-1} \exp V(A \cup B),$$

with
$$Z_{\Lambda,B} = \sum_{E \subset \Lambda} \exp V(E \cup B). \square$$

If $\mathscr{V} = \{f^{\Lambda}\}_{\Lambda \in \mathscr{C}(S)}$ is a positive local specification and V is the potential given by Proposition 5.3 then we will call V the *potential associated with* $\mathscr{V}$. We will also write $\mathscr{G}_V$ instead of $\mathscr{G}(\mathscr{V})$ and call $\mathscr{G}_V$ the set of *Gibbs states with potential* V. A potential V on S will be called *continuous* if given any $A \in \mathscr{C}(S)$ and $\epsilon > 0$ then there exists $\Lambda \in \mathscr{C}(S-\Lambda)$ such that

$$\text{if} \quad B_1, B_2 \in \mathscr{C}(S-A) \quad \text{with} \quad B_1 \supset \Lambda, B_2 \supset \Lambda$$

$$\text{then} \quad |(V(A \cup B_1) - V(B_1)) - (V(A \cup B_2) - V(B_2))| < \epsilon.$$

Of course, this just says that given $A \in \mathscr{C}(S)$ then there exists $g \in C(\mathscr{P}(S-A))$ such that

$$V(A \cup B) - V(B) = g(B) \quad \text{for all} \quad B \in \mathscr{C}(S-A).$$

(g is uniquely defined since $\mathscr{C}(S-A)$ is dense in $\mathscr{P}(S-A)$.) If V is the potential associated with $\mathscr{V}$ then V is a continuous potential since

$$V(A \cup B) - V(B) = \log\left[\frac{f^{A}(A,B)}{f^{A}(\varnothing, B)}\right] \quad \text{for all} \quad B \in \mathscr{C}(S-A).$$

The converse of this is true:

PROPOSITION 5.4 Let V be a continuous potential on S. Then there exists a unique positive local specification

$$\mathscr{V} = \{f^{\Lambda}\}_{\Lambda \in \mathscr{C}(S)}$$

such that V is the potential associated with $\mathscr{V}$.

Proof If $A \subset \Lambda \in \mathscr{C}(S)$ and $B \in \mathscr{C}(S-\Lambda)$ then we have to define

$$f^{\Lambda}(A,B) = Z_{\Lambda,B}^{-1} \exp V(A \cup B)$$

with

$$Z_{\Lambda,B} = \sum_{E \subset \Lambda} \exp V(E \cup B).$$

Now we can write

$$f^{\Lambda}(A,B) = \frac{\exp[V(A \cup B) - V(B)]}{\sum_{E \subset \Lambda} \exp[V(E \cup B) - V(B)]}$$

and V is a continuous potential; thus $f^{\Lambda}(A, \cdot)$ extends uniquely to a continuous function on $\mathscr{P}(S-\Lambda)$. It is now a simple matter to check that $\{f^{\Lambda}\}_{\Lambda \in \mathscr{C}(S)}$ is a positive local specification. □

If V is a potential on S then we define $J_V : \mathscr{C}(S) \to \mathbb{R}$ by

$$J_V(A) = \sum_{X \subset A} (-1)^{|A-X|} V(X).$$

J_V is again a potential on S and we will call it the *interaction potential* corresponding to V. It is easily checked that we have

$$V(A) = \sum_{B \subset A} J_V(B) \quad \text{for all} \quad A \in \mathscr{C}(S).$$

Conversely, given a potential Φ on S let us denote by U_{Φ} the potential given by

$$U_{\Phi}(A) = \sum_{B \subset A} \Phi(B).$$

Thus Φ is the interaction potential corresponding to U_{Φ} and we have

$$\Phi(A) = \sum_{X \subset A} (-1)^{|A-X|} U_{\Phi}(X).$$

We let $D(S)$ denote the set of local specifications on S and $D_+(S)$ the set of positive local specifications. Also let $H(S)$ denote the set of continuous potentials on S. We can consider $H(S)$ as a real vector space and by Propositions 5.3 and 5.4 we can identify $H(S)$ and $D_+(S)$.

Let $\mathscr{V} \in D(S)$ (resp. $V \in H(S)$); if $\mathscr{G}(\mathscr{V})$ (resp. $\mathscr{G}_V$) consists of more than one state then we will say that *phase transition* occurs for $\mathscr{V}$ (resp. V). Most of the rest of the book is taken up with finding out conditions under which phase transition can or cannot occur.

If $\mu \in \mathscr{S}(S)$ then we define the *correlation function* of μ to be the function $\rho: \mathscr{C}(S) \to \mathbb{R}$ given by

$$\rho(A) = \mu([A, A]) \quad \text{for} \quad A \in \mathscr{C}(S).$$

Thus $\rho(A) = \mu(\{X \in \mathscr{P}(S): X \supset A\})$. If $A \subset \Lambda \in \mathscr{C}(S)$ then we have

$$\mu([A, \Lambda]) = \sum_{E \subset \Lambda - A} (-1)^{|E|} \rho(A \cup E)$$

(where $|E|$ denotes the number of points in E). Therefore ρ determines $\mu([A, \Lambda])$ for every finite dimensional cylinder $[A, \Lambda]$ and hence ρ determines μ. We will usually find it easier to work with ρ than to work with μ.

Let $\mathscr{V} \in D(S)$ with $\mathscr{V} = \{f^{\Lambda}\}_{\Lambda \in \mathscr{C}(S)}$; we will consider $\mathscr{V}$ to be fixed from now on and will thus define things that depend on $\mathscr{V}$ using notation that does not mention $\mathscr{V}$ (and hope that no confusion arises). For example, if $\Lambda \in \mathscr{C}(S)$ and $X \in \mathscr{P}(S - \Lambda)$ then we define $\pi_{\Lambda}^{X} \in \mathscr{S}(\Lambda)$ by

$$\pi_{\Lambda}^{X}([A, \Lambda]) = f^{\Lambda}(A, X) \quad \text{for} \quad A \subset \Lambda.$$

We will call π_{Λ}^{X} the *Gibbs state on* Λ *(with specification* $\mathscr{V}$*) and boundary values* X. Let ρ_{Λ}^{X} denote the correlation function of π_{Λ}^{X}, thus $\rho_{\Lambda}^{X}: \mathscr{C}(\Lambda) \to \mathbb{R}$ is defined by

$$\rho_{\Lambda}^{X}(A) = \sum_{A \subset B \subset \Lambda} f^{\Lambda}(B, X) \quad \text{for} \quad A \subset \Lambda.$$

PROPOSITION 5.5 Let $\mu \in \mathscr{G}(\mathscr{V})$ with correlation function ρ. We have:

(i) If $A \subset B \subset \Lambda \in \mathscr{C}(S)$ then

$$\mu([A, B]) = \int \pi_{\Lambda}^{X}([A, B]) \, \mathrm{d}r_{S-\Lambda}(\mu)(X).$$

(ii) If $A \subset \Lambda \in \mathscr{C}(S)$ then

$$\rho(A) = \int \rho_{\Lambda}^{X}(A) \, \mathrm{d}r_{S-\Lambda}(\mu)(X).$$

(iii) If $A \subset \Lambda \in \mathscr{C}(S)$ then

$$\min_{X \in \mathscr{P}(S-\Lambda)} \rho_{\Lambda}^{X}(A) \leqslant \rho(A) \leqslant \max_{X \in \mathscr{P}(S-\Lambda)} \rho_{\Lambda}^{X}(A).$$

Proof It is clear that (i) $\Rightarrow$ (ii) $\Rightarrow$ (iii) and to check that (i) holds is an easy calculation. $\square$

Part (iii) of Proposition 5.5 will be used to obtain bounds on the correlation functions of Gibbs states. If $\Lambda \in \mathscr{C}(S)$ then we define $\rho_\Lambda^+ : \mathscr{C}(\Lambda) \to \mathbb{R}, \rho_\Lambda^- : \mathscr{C}(\Lambda) \to \mathbb{R}$ by

$$\rho_\Lambda^+(A) = \max_{X \in \mathscr{P}(S-\Lambda)} \rho_\Lambda^X(A),$$

$$\rho_\Lambda^-(A) = \min_{X \in \mathscr{P}(S-\Lambda)} \rho_\Lambda^X(A).$$

[Note that in general ρ_Λ^+ and ρ_Λ^- are not correlation functions since the $X \in \mathscr{P}(S-\Lambda)$, for which the maximum or minimum is obtained, depends on A.]

PROPOSITION 5.6 $\rho_\Lambda^+(A)$ is a decreasing function of Λ, i.e. if $A \subset \Lambda \subset \Lambda' \in \mathscr{C}(S)$ then $\rho_{\Lambda'}^+(A) \leqslant \rho_\Lambda^+(A)$. Similarly $\rho_\Lambda^-(A)$ is an increasing function of Λ.

Proof Let $A \subset \Lambda \in \mathscr{C}(S)$ and let $x \notin \Lambda$. Clearly we need only prove that $\rho_{\Lambda\cup x}^+(A) \leqslant \rho_\Lambda^+(A)$ and $\rho_{\Lambda\cup x}^-(A) \geqslant \rho_\Lambda^-(A)$. If

$$X \in \mathscr{C}(S-(\Lambda \cup x))$$

then we have

$$\begin{aligned}
&\rho_{\Lambda\cup x}^X(A)\\
&\quad = \sum_{A \subset B \subset \Lambda \cup x} f^{\Lambda\cup x}(B, X)\\
&\quad = \sum_{A \subset B \subset \Lambda} [f^{\Lambda\cup x}(B, X) + f^{\Lambda\cup x}(B \cup x, X)]\\
&\quad = \sum_{A \subset B \subset \Lambda} [f^{\Lambda}(B, X) \sum_{C \subset \Lambda} f^{\Lambda\cup x}(C, X) + f^{\Lambda}(B, X \cup x) \sum_{C \subset \Lambda} f^{\Lambda\cup x}(C \cup x, X)]\\
&\quad = \rho_\Lambda^X(A) \sum_{C \subset \Lambda} f^{\Lambda\cup x}(C, X) + \rho_\Lambda^{X\cup x}(A) \sum_{C \subset \Lambda} f^{\Lambda\cup x}(C \cup x, X).
\end{aligned}$$

Therefore

$$\begin{aligned}
\rho_{\Lambda\cup x}^X(A) &\leqslant \rho_\Lambda^+(A) [\sum_{C \subset \Lambda} f^{\Lambda\cup x}(C, X) + \sum_{C \subset \Lambda} f^{\Lambda\cup x}(C \cup x, X)]\\
&= \rho_\Lambda^+(A) \sum_{C \subset \Lambda\cup x} f^{\Lambda\cup x}(C, X) = \rho_\Lambda^+(A),
\end{aligned}$$

and thus $\rho_{\Lambda\cup x}^+(A) \leqslant \rho_\Lambda^+(A)$. In exactly the same way we have $\rho_{\Lambda\cup x}^-(A) \geqslant \rho_\Lambda^-(A)$. $\square$

We now define $\rho^+ : \mathscr{C}(S) \to \mathbb{R}$, $\rho^- : \mathscr{C}(S) \to \mathbb{R}$ by

$$\rho^+(A) = \lim_{\Lambda \uparrow S} \rho^+_\Lambda(A),$$

$$\rho^-(A) = \lim_{\Lambda \uparrow S} \rho^-_\Lambda(A),$$

and of course by Proposition 5.6 these limits exist. If ρ is the correlation function of $\mu \in \mathscr{G}(\mathscr{V})$ then by Proposition 5.5 we have

$$\rho^-(A) \leqslant \rho(A) \leqslant \rho^+(A) \quad \text{for all} \quad A \in \mathscr{C}(S).$$

Therefore if $\rho^- = \rho^+$ then phase transition cannot occur for $\mathscr{V}$. The converse of this is true and in order to prove it we need:

LEMMA 5.1 Let $\Lambda_n \in \mathscr{C}(S)$ with $\Lambda_n \uparrow S$ and let $X_n \in \mathscr{P}(S - \Lambda_n)$. Define $\mu_n \in \mathscr{S}(S)$ by

$$\mu_n([A, \Lambda_n]) = f^{\Lambda_n}(A, X_n) \quad \text{for all} \quad A \subset \Lambda_n,$$

$$\mu_n([x, x]) = 1 \quad \text{for all} \quad x \in X_n,$$

$$\mu_n([\varnothing, x]) = 1 \quad \text{for all} \quad x \notin \Lambda_n \cup X_n.$$

Suppose that $\mu_n \to \mu$ as $n \to \infty$. Then $\mu \in \mathscr{G}(\mathscr{V})$.

Proof The proof is almost exactly the same as the proof of Proposition 5.2 and is thus left for the reader. □

PROPOSITION 5.7 If $A \in \mathscr{C}(S)$ then there exist $\mu_1, \mu_2 \in \mathscr{G}(\mathscr{V})$ such that if ρ_1 (resp. ρ_2) is the correlation function of μ_1 (resp. μ_2) then $\rho_1(A) = \rho^+(A)$ and $\rho_2(A) = \rho^-(A)$.

Proof Let $A \in \mathscr{C}(S)$ and choose $\Lambda_n \in \mathscr{C}(S)$ with $\Lambda_n \supset A$ and $\Lambda_n \uparrow S$. Let $X_n \in \mathscr{P}(S - \Lambda_n)$ be such that $\rho^{X_n}_{\Lambda_n}(A) = \rho^+_{\Lambda_n}(A)$ and let $\mu_n \in \mathscr{S}(S)$ be defined as in Lemma 5.1. By choosing a subsequence we can assume that there exists $\mu \in \mathscr{S}(S)$ such that $\mu_n \to \mu$ as $n \to \infty$. Therefore from Lemma 5.1 we have $\mu \in \mathscr{G}(\mathscr{V})$. Let ρ be the correlation function of μ; then

$$\rho(A) = \lim_{n \to \infty} \mu_n([A, A]) = \lim_{n \to \infty} \rho^{X_n}_{\Lambda_n}(A)$$

$$= \lim_{n \to \infty} \rho^+_{\Lambda_n}(A) = \rho^+(A).$$

The other half of the Proposition follows in the same way. □

Combining Propositions 5.5 and 5.7 immediately gives:

THEOREM 5.1 If $\mathscr{V} \in D(S)$ then phase transition occurs for $\mathscr{V}$ if and only if $\rho^+ \neq \rho^-$.

NOTES As mentioned in Chapter 4 the definition of a Gibbs state used here is that given by Dobrushin (1968b, 1968c, 1969).

6. *Kirkwood—Salsburg equations*

Let $V \in H(S)$ and let $\mu \in \mathscr{G}_V$ with correlation function ρ. We will show that ρ satisfies a set of equations called the Kirkwood–Salsburg equations. For some potentials V we will be able to prove that the Kirkwood–Salsburg equations have at most one solution, and thus phase transition cannot occur for these potentials. The proof of the Kirkwood–Salsburg equations depends on the following partial summation formula.

LEMMA 6.1 Let $\Lambda \in \mathscr{C}(S)$ and $f, g: \mathscr{P}(\Lambda) \to \mathbb{R}$. Then

$$\sum_{A \subset \Lambda} f(A)g(A) = \sum_{A \subset \Lambda} \Big[\sum_{X \subset A} (-1)^{|A-X|} f(X)\Big]\Big[\sum_{A \subset Y \subset \Lambda} g(Y)\Big].$$

Proof Define

$$\tau_{X,Y}(A) = \begin{cases} 1 & \text{if } X \subset A \subset Y, \\ 0 & \text{otherwise.} \end{cases}$$

Then

$$\begin{aligned} &\sum_{A \subset \Lambda} \Big[\sum_{X \subset A} (-1)^{|A-X|} f(X)\Big]\Big[\sum_{A \subset Y \subset \Lambda} g(Y)\Big] \\ &= \sum_{A \subset \Lambda} \sum_{X \subset \Lambda} \sum_{Y \subset \Lambda} (-1)^{|A-X|} f(X) g(Y) \tau_{X,Y}(A) \\ &= \sum_{X \subset \Lambda} \sum_{Y \subset \Lambda} f(X) g(Y) \sum_{X \subset A \subset Y} (-1)^{|A-X|} \\ &= \sum_{X \subset \Lambda} f(X) g(X), \end{aligned}$$

since $\sum_{X \subset A \subset Y} (-1)^{|A-X|} = 1$ if $X = Y$ and 0 otherwise. □

Let $\mathscr{V} \in D_+(S)$ with $\mathscr{V} = \{f^\Lambda\}_{\Lambda \in \mathscr{C}(S)}$, and let $V \in H(S)$ be the potential associated with $\mathscr{V}$. We now have:

THEOREM 6.1 Let $x \in \Lambda \in \mathscr{C}(S)$ and let $\mu \in \mathscr{G}(\mathscr{V})$ with correlation function ρ. Then

$$\rho(\Lambda) = \lim_{\Lambda' \to S} \sum_{B \subset \Lambda' - \Lambda} \Big[\sum_{X \subset B} (-1)^{|B-X|} \exp h(X)\Big] \times [\rho((\Lambda - x) \cup B) - \rho(\Lambda \cup B)],$$

where $h \in C(\mathscr{P}(S-\Lambda))$ is given by

$$h(X) = V(\Lambda \cup X) - V((\Lambda - x) \cup X) \quad \text{for} \quad X \in \mathscr{C}(S-\Lambda).$$

Proof We have

$$\begin{aligned}\rho(\Lambda) &= \int \mathrm{d}r^{\Lambda}_{S-\Lambda}(\mu) = \int f^{\Lambda}(\Lambda, X)\,\mathrm{d}r_{S-\Lambda}(\mu)\,(X)\\ &= \int \exp h(X) f^{\Lambda}(\Lambda - x, X)\,\mathrm{d}r_{S-\Lambda}(\mu)\,(X)\\ &= \int \exp h(X)\,\mathrm{d}r^{\Lambda-x}_{S-\Lambda}(\mu)\,(X).\end{aligned}$$

But $h \in C(\mathscr{P}(S-\Lambda))$, thus we have

$$\begin{aligned}\rho(\Lambda) &= \lim_{\Lambda' \to S} \sum_{B \subset \Lambda' - \Lambda} \exp h(B)\, r^{\Lambda-x}_{S-\Lambda}(\mu)\,([B, \Lambda' - \Lambda])\\ &= \lim_{\Lambda' \to S} \sum_{B \subset \Lambda' - \Lambda} \exp h(B)\, \mu([(\Lambda - x) \cup B, \Lambda']).\end{aligned}$$

Now it is easily checked that

$$\sum_{B \subset Y \subset \Lambda' - \Lambda} \mu([(\Lambda - x) \cup B, \Lambda']) = \rho((\Lambda - x) \cup B) - \rho(\Lambda \cup B).$$

Therefore by Lemma 6.1 we have

$$\rho(\Lambda) = \lim_{\Lambda' \to S} \sum_{B \subset \Lambda' - \Lambda} \Big[\sum_{X \subset B} (-1)^{|B-X|} \exp h(X)\Big] \times [\rho((\Lambda - x) \cup B) - \rho(\Lambda \cup B)]. \square$$

If $x \in \Lambda \in \mathscr{C}(S)$ then we define $K_x(\Lambda, \cdot\,) : \mathscr{C}(S - (\Lambda - x)) \to \mathbb{R}$ by

$$K_x(\Lambda, B) = \sum_{X \subset B - x} (-1)^{|B-X|} \exp[V(\Lambda \cup X) - V((\Lambda - x) \cup X)].$$

Then Theorem 6.1 gives that

$$\rho(\Lambda) = \lim_{\Lambda' \to S} \sum_{B \subset \Lambda' - (\Lambda - x)} K_x(\Lambda, B)\, \rho((\Lambda - x) \cup B).$$

These equations (for all $x \in \Lambda \in \mathscr{C}(S)$) are called the *Kirkwood–Salsburg equations.*

PROPOSITION 6.1 Suppose there exists $\alpha < 1$ such that for all $x \in \Lambda \in \mathscr{C}(S)$ we have

$$\lim_{\Lambda' \to S} \sum_{B \subset \Lambda' - (\Lambda - x)} |K_x(\Lambda, B)| \leqslant \alpha.$$

Then phase transition cannot occur for V.

Proof Let $\mu_1, \mu_2 \in \mathscr{G}_V$ with correlation functions ρ_1, ρ_2 and let

$$m = \sup_{A \in \mathscr{C}(S)} |\rho_1(A) - \rho_2(A)|.$$

(Thus $0 \leqslant m \leqslant 1$.) Take $\Lambda \in \mathscr{C}(S)$ with $\Lambda \neq \varnothing$, and let $x \in \Lambda$. Then

$$\rho_1(\Lambda) - \rho_2(\Lambda) = \lim_{\Lambda' \to S} \sum_{B \subset \Lambda' - (\Lambda - x)} K_x(\Lambda, B)\,(\rho_1((\Lambda - x) \cup B) - \rho_2((\Lambda - x) \cup B))$$

and thus

$$|\rho_1(\Lambda) - \rho_2(\Lambda)| \leqslant \lim_{\Lambda' \to S} \sum_{B \subset \Lambda' - (\Lambda - x)} |K_x(\Lambda, B)|\, m \leqslant \alpha m.$$

Therefore
$$\sup_{\Lambda \in \mathscr{C}(S),\, \Lambda \neq \varnothing} |\rho_1(\Lambda) - \rho_2(\Lambda)| \leqslant \alpha m.$$

But $\rho_1(\varnothing) = \rho_2(\varnothing) = 1$ and thus

$$m = \sup_{\Lambda \in \mathscr{C}(S),\, \Lambda \neq \varnothing} |\rho_1(\Lambda) - \rho_2(\Lambda)| \leqslant \alpha m.$$

Hence $m = 0$, i.e. $\rho_1 = \rho_2$, and therefore phase transition cannot occur for V. □

Of course, what we have to do now is to find conditions on the potential V under which the hypotheses of Proposition 6.1 hold. To this end the following simple lemma is very useful.

LEMMA 6.2 Let $J: \mathscr{C}(S) \to \mathbb{R}$ and define $W: \mathscr{C}(S) \to \mathbb{R}$ by

$$W(A) = \sum_{X \subset A} J(X) \quad \text{for} \quad A \in \mathscr{C}(S).$$

Then for any $B \in \mathscr{C}(S)$ with $B \neq \varnothing$

$$\sum_{X \subset B} (-1)^{|B - X|} \exp W(X)$$

$$= \sum_{n=1}^{\infty} \frac{1}{n!} \left[\sum_{\substack{Y_1, \ldots, Y_n \subset B \\ \text{such that } Y_1 \cup \ldots \cup Y_n = B}} J(Y_1) \ldots J(Y_n) \right].$$

Proof We have

$$\sum_{X\subset B}(-1)^{|B-X|}\exp W(X) = \sum_{X\subset B}(-1)^{|B-X|}\exp[\sum_{Y\subset X}J(Y)]$$

$$= \sum_{X\subset B}(-1)^{|B-X|}\sum_{n=0}^{\infty}\frac{1}{n!}[\sum_{Y_1\subset X}\cdots\sum_{Y_n\subset X}J(Y_1)\dots J(Y_n)]$$

$$= \sum_{n=1}^{\infty}\frac{1}{n!}\sum_{Y_1\subset B}\cdots\sum_{Y_n\subset B}J(Y_1)\dots J(Y_n)\sum_{Y_1\cup\dots\cup Y_n\subset X\subset B}(-1)^{|B-X|}$$

$$= \sum_{n=1}^{\infty}\frac{1}{n!}\left[\sum_{\substack{Y_1\dots Y_n\subset B\\ \text{such that } Y_1\cup\dots\cup Y_n=B}}J(Y_1)\dots J(Y_n)\right].\ \square$$

Immediate consequences of the lemma are firstly, that if $J(X) \geqslant 0$ for all $X\in\mathscr{C}(S)$ then

$$\sum_{X\subset B}(-1)^{|B-X|}\exp[\sum_{Y\subset X}J(Y)] \geqslant 0 \quad \text{for all} \quad B\in\mathscr{C}(S);$$

and secondly that for any $J:\mathscr{C}(S)\to\mathbb{R}$ we have

$$|\sum_{X\subset B}(-1)^{|B-X|}\exp[\sum_{Y\subset X}J(Y)]| \leqslant \sum_{X\subset B}(-1)^{|B-X|}\exp[\sum_{Y\subset X}|J(Y)|].$$

THEOREM 6.2 Suppose there exists $\alpha < 1$ such that

$$J_V(\{x\}) + \sum_{\substack{\varnothing\neq A\in\mathscr{C}(S)\\ x\notin A}}|J_V(A\cup x)| \leqslant \log\left(\frac{\alpha}{2}\right)$$

for all $x\in S$. Then phase transition cannot occur for V.

Proof Let $x\in\Lambda\subset\Lambda'\in\mathscr{C}(S)$ and $B\subset\Lambda'-\Lambda$. If $X\subset B$ then

$$V(\Lambda\cup X)-V((\Lambda-x)\cup X) = \sum_{Y\subset\Lambda\cup X}J_V(Y) - \sum_{Y\subset(\Lambda-x)\cup X}J_V(Y)$$

$$= \sum_{Y\subset(\Lambda-x)\cup X}J_V(Y\cup x) = \sum_{Y\subset X}\sum_{E\subset(\Lambda-x)}J_V(E\cup x\cup Y)$$

$$= \sum_{E\subset(\Lambda-x)}J_V(E\cup x) + \sum_{Y\subset X}J(Y),$$

where
$$J(Y) = \begin{cases}\sum_{E\subset(\Lambda-x)}J_V(E\cup x\cup Y) & \text{if} \quad Y\neq\varnothing,\\ 0 & \text{if} \quad Y=\varnothing.\end{cases}$$

Therefore

$$\Big|\sum_{X\subset B}(-1)^{|B-X|}\exp\left[V(\Lambda\cup X)-V((\Lambda-x)\cup X)\right]\Big|$$

$$=\exp\Big[\sum_{E\subset(\Lambda-x)}J_V(E\cup x)\Big]\Big|\sum_{X\subset B}(-1)^{|B-X|}\exp\Big(\sum_{Y\subset X}J(Y)\Big)\Big|,$$

and by Lemma 6.2 we have

$$\Big|\sum_{X\subset B}(-1)^{|B-X|}\exp\Big(\sum_{Y\subset X}J(Y)\Big)\Big|\leqslant\sum_{X\subset B}(-1)^{|B-X|}\exp\Big(\sum_{Y\subset X}|J|(Y)\Big),$$

where $$|J|(Y)=\begin{cases}\sum_{E\subset(\Lambda-x)}|J_V(E\cup x\cup Y)| & \text{if } Y\neq\varnothing,\\ 0 & \text{if } Y=\varnothing.\end{cases}$$

Thus

$$|K_x(\Lambda,B)|\leqslant\exp\Big[\sum_{E\subset(\Lambda-x)}J_V(E\cup x)\Big]$$

$$\times\sum_{X\subset B}(-1)^{|B-X|}\exp\Big(\sum_{Y\subset X}|J|(Y)\Big)$$

$$\leqslant\sum_{X\subset B}(-1)^{|B-X|}\exp\Big[\sum_{\varnothing\neq Y\subset(\Lambda-x)\cup X}|J_V(Y\cup x)|+J_V(\{x\})\Big].$$

Note that if $W:\mathscr{C}(S)\to\mathbb{R}$ is any function then

$$\sum_{B\subset\Lambda'-\Lambda}\ \sum_{X\subset B}(-1)^{|B-X|}\exp W(X)$$

$$=\sum_{X\subset\Lambda'-\Lambda}\exp W(X)\sum_{X\subset B\subset\Lambda'-\Lambda}(-1)^{|B-X|}=\exp W(\Lambda'-\Lambda).$$

Therefore we have

$$\sum_{B\subset\Lambda'-\Lambda}|K_x(\Lambda,B)|\leqslant\exp\Big[J_V(\{x\})+\sum_{\varnothing\neq Y\subset\Lambda'-x}|J_V(Y\cup x)|\Big]$$

$$\leqslant\exp\Big[J_V(\{x\})+\sum_{\substack{\varnothing\neq A\in\mathscr{C}(S)\\ x\notin A}}|J_V(A\cup x)|\Big].$$

Thus $$\sum_{B\subset\Lambda'-(\Lambda-x)}|K_x(\Lambda,B)|=2\sum_{B\subset\Lambda'-\Lambda}|K_x(\Lambda,B)|$$

$$\leqslant 2\exp\Big[J_V(\{x\})+\sum_{\substack{\varnothing\neq A\in\mathscr{C}(S)\\ x\notin A}}|J_V(A\cup x)|\Big]\leqslant\alpha,$$

and hence by Proposition 6.1 phase transition cannot occur for V. □

Note that for Theorem 6.2 to work we need $J_V(\{x\})$ negative for all $x \in S$ (and in fact we must have $J_V(\{x\}) < -\log 2$ for all $x \in S$). In the next chapter we will show how the theorem can be adapted to work when $J_V(\{x\})$ is positive. The situations in which Theorem 6.2 works correspond to situations in classical statistical mechanics where phase transition does not occur. For example

$$\sum_{\substack{\varnothing \neq A \in \mathscr{C}(S) \\ x \notin A}} |J_V(A \cup x)|$$

being small for all $x \in S$ means that the interactions in the physical system are very weak. If $J_V(\{x\}) < 0$ and $|J_V(\{x\})|$ is large for all $x \in S$ then this corresponds to a strong external field being applied to the system. Finally if we replace V by $t^{-1}V$ (for $t > 0$) then the parameter t is proportional to the temperature of the system, and thus Theorem 6.2 tells us that we should not have phase transition at high temperatures.

NOTES The fact that the correlation function of a Gibbs state must satisfy equations like the Kirkwood–Salsburg equations was first noted by Mayer (1947). The techniques used here are adapted from the development in Lanford and Ruelle (1969) for pair potentials.

7. *Involutions of* $\mathscr{P}(S)$

The object of our study in the last two chapters has been certain probability measures on $\mathscr{P}(S)$. We have the obvious interpretation that $X \in \mathscr{P}(S)$ describes when there are particles (or some other entity) at the points of X and no particles at the points of $S-X$. When we put the topology on $\mathscr{P}(S)$ we regarded $\mathscr{P}(S)$ as $\{0, 1\}^S$ and in $\{0, 1\}^S$ there is no real difference between 0 and 1 as they just act as labels. We will now consider what happens if we interchange 0 and 1 for certain points of S. Let $R \subset S$; we will interchange 0 and 1 for all $x \in R$, and thus consider a transformation $\tau_R: \mathscr{P}(S) \to \mathscr{P}(S)$ given by

$$\tau_R(A) = (A \cap (S-R)) \cup (R-A) \quad \text{for all} \quad A \in \mathscr{P}(S).$$

It is clear that τ_R is an involution (i.e. $\tau_R(\tau_R(A)) = A$ for all $A \in \mathscr{P}(S)$) and thus the pretentious title of this chapter. We will mainly be interested in the case $R = S$, i.e. when all the 0's and 1's are interchanged. Note that

$$\tau_R([A, \Lambda]) = [\tau_R(A) \cap \Lambda, \Lambda]$$

and thus τ_R is a homeomorphism. We let τ_R induce a map

$$\tau_R: \mathscr{M}(S) \to \mathscr{M}(S)$$

by $$\tau_R(\mu)(\Omega) = \mu(\tau_R(\Omega)) \quad \text{for all} \quad \Omega \in \mathscr{F}(S), \mu \in \mathscr{M}(S).$$

It is easy to see that τ_R is a homeomorphism from $\mathscr{M}(S)$ to $\mathscr{M}(S)$ and maps $\mathscr{S}(S)$ onto $\mathscr{S}(S)$; again τ_R is an involution.

If $A \in \mathscr{P}(S)$ then we define $\tau_R^A: \mathscr{P}(A) \to \mathscr{P}(A)$ by

$$\tau_R^A(B) = \tau_R(B) \cap A, \quad \text{for all} \quad B \in \mathscr{P}(A).$$

Then τ_R^A is a homeomorphism from $\mathscr{P}(A)$ to $\mathscr{P}(A)$, and if $A_1, A_2 \in \mathscr{P}(S)$ with $A_1 \cap A_2 = \varnothing$ and $B_1 \subset A_1$, $B_2 \subset A_2$ then

$$\tau_R^{A_1 \cup A_2}(B_1 \cup B_2) = \tau_R^{A_1}(B_1) \cup \tau_R^{A_2}(B_2).$$

PROPOSITION 7.1 Let $\mathscr{V} \in D(S)$ with $\mathscr{V} = \{f^\Lambda\}_{\Lambda \in \mathscr{C}(S)}$; let $R \in \mathscr{P}(S)$ and define $\mathscr{V}_R = \{g^\Lambda\}_{\Lambda \in \mathscr{C}(S)}$ by

$$g^\Lambda(A, X) = f^\Lambda(\tau_R^\Lambda(A), \tau_R^{S-\Lambda}(X)) \quad \text{for} \quad A \in \mathscr{P}(\Lambda), X \in \mathscr{P}(S-\Lambda).$$

Then $\mathscr{V}_R \in D(S)$ and if $\mu \in \mathscr{G}(\mathscr{V})$ then $\tau_R(\mu) \in \mathscr{G}(\mathscr{V}_R)$.

Proof It is a simple matter to check that

$$g^\Lambda \in C(\mathscr{P}(\Lambda) \times \mathscr{P}(S-\Lambda))$$

and of course $g^\Lambda(A, X) \geqslant 0$ for all $A \in \mathscr{P}(\Lambda), X \in \mathscr{P}(S-\Lambda)$. Also we must have

$$\sum_{A \subset \Lambda} g^\Lambda(A, X) = 1 \quad \text{for all} \quad X \in \mathscr{P}(S-\Lambda),$$

since τ_R^Λ is just a permutation of $\mathscr{P}(\Lambda)$. Now let $\Lambda \subset \Lambda' \in \mathscr{C}(S)$, $A \subset \Lambda$, $B \subset \Lambda' - \Lambda$ and $X \subset S - \Lambda'$. Then

$$\begin{aligned} g^{\Lambda'}(A \cup B, X) &= f^{\Lambda'}(\tau_R^{\Lambda'}(A \cup B), \tau_R^{S-\Lambda'}(X)) \\ &= f^{\Lambda'}(\tau_R^\Lambda(A) \cup \tau_R^{\Lambda'-\Lambda}(B), \tau_R^{S-\Lambda'}(X)) \\ &= f^\Lambda(\tau_R^\Lambda(A), \tau_R^{\Lambda'-\Lambda}(B) \cup \tau_R^{S-\Lambda'}(X)) \sum_{C \subset \Lambda} f^{\Lambda'}(C \cup \tau_R^{\Lambda'-\Lambda}(B), \tau_R^{S-\Lambda'}(X)) \\ &= f^\Lambda(\tau_R^\Lambda(A), \tau_R^{S-\Lambda}(B \cup X)) \sum_{C \subset \Lambda} f^{\Lambda'}(\tau_R^\Lambda(C) \cup \tau_R^{\Lambda'-\Lambda}(B), \tau_R^{S-\Lambda'}(X)) \\ &= g^\Lambda(A, B \cup X) \sum_{C \subset \Lambda} g^{\Lambda'}(C \cup B, X). \end{aligned}$$

Therefore $\mathscr{V}_R \in D(S)$. Now let $\mu \in \mathscr{G}(\mathscr{V})$ and $A \subset \Lambda \in \mathscr{C}(S)$, $E \subset F \in \mathscr{C}(S-\Lambda)$. Then

$$\begin{aligned} & r_{S-\Lambda}^A(\tau_R(\mu))([E, F])\, \tau_R(\mu)([A \cup E, \Lambda \cup F]) \\ &= \mu([\tau_R^\Lambda(A) \cup \tau_R^F(E), \Lambda \cup F]) \\ &= r_{S-\Lambda}^{\tau_R^\Lambda(A)}(\mu)([\tau_R^F(E), F]) \\ &= \int_{[\tau_R^F(E), F]} f^\Lambda(\tau_R^\Lambda(A), X)\, \mathrm{d}r_{S-\Lambda}(\mu)(X) \\ &= \int_{[E,F]} f^\Lambda(\tau_R^\Lambda(A), \tau_R^{S-\Lambda}(Y))\, \mathrm{d}r_{S-\Lambda}(\mu)(\tau_R^{S-\Lambda}(Y)) \\ &= \int_{[E,F]} f^\Lambda(\tau_R^\Lambda(A), \tau_R^{S-\Lambda}(Y))\, \mathrm{d}r_{S-\Lambda}(\tau_R(\mu))(Y) \\ &= \int_{[E,F]} g^\Lambda(A, Y)\, \mathrm{d}r_{S-\Lambda}(\tau_R(\mu))(Y). \end{aligned}$$

Thus $r^A_{S-\Lambda}(\tau_R(\mu)) = g^\Lambda(A, \cdot\,)\, r_{S-\Lambda}(\tau_R(\mu))$, and hence

$$\tau_R(\mu) \in \mathscr{G}(\mathscr{V}_R).\square$$

Now since $\tau_R\colon \mathscr{M}(S) \to \mathscr{M}(S)$ is a homeomorphism it follows immediately from Proposition 7.1 that τ_R maps $\mathscr{G}(\mathscr{V})$ homeomorphically onto $\mathscr{G}(\mathscr{V}_R)$. In particular, phase transition occurs for $\mathscr{V}$ if and only if it occurs for $\mathscr{V}_R$. If $\mathscr{V} \in D_+(S)$ then it is clear that $\mathscr{V}_R \in D_+(S)$. Thus suppose that $\mathscr{V} \in D_+(S)$ and let V be the potential associated with $\mathscr{V}$, also let V_R be the potential associated with $\mathscr{V}_R$. We will now find out what V_R is in terms of V.

PROPOSITION 7.2 $V_R\colon \mathscr{C}(S) \to \mathbb{R}$ is given by

$$V_R(A) = V((A-R) \cup (R-A)) - V(R)$$

for any $A \in \mathscr{C}(S)$. (Note that $V((A-R) \cup (R-A)) - V(R)$ makes sense since V is a continuous potential.)

Proof Let $\mathscr{V}_R = \{g^\Lambda\}_{\Lambda \in \mathscr{C}(S)}$ as before. Then we must have for any $A \in \mathscr{C}(S)$,

$$V_R(A) = \log\left[\frac{g^A(A, \varnothing)}{g^A(\varnothing, \varnothing)}\right]$$

$$= \log\left[\frac{f^A(\tau^A_R(A), \tau^{S-A}_R(\varnothing))}{f^A(\tau^A_R(\varnothing), \tau^{S-A}_R(\varnothing))}\right] = \log\left[\frac{f^A(A-R, R-A)}{f^A(A, R-A)}\right],$$

and it is quite easy to check that

$$\log\left[\frac{f^A(A-R, R-A)}{f^A(A, R-A)}\right] = V((A-R) \cup (R-A)) - V(R).$$

Thus $$V_R(A) = V((A-R) \cup (R-A)) - V(R).\square$$

From now on we will only consider the case when $R = S$. For convenience we will write $\bar{V}$ instead of V_S. By Proposition 7.2 we have

$$\bar{V}(A) = V(S-A) - V(S) \quad \text{for all} \quad A \in \mathscr{C}(S).$$

PROPOSITION 7.3 If $V \in H(S)$ then

$$J_{\bar{V}}(A) = (-1)^{|A|} \sum_{\substack{Y \in \mathscr{C}(S) \\ Y \supset A}} J_V(Y) \quad \text{for all} \quad A \in \mathscr{C}(S), A \neq \varnothing.$$

(Of course, for $A = \varnothing$ we have $J_{\bar{V}}(A) = 0$.)

Proof By $\sum_{\substack{Y \in \mathscr{C}(S)\\ Y \supset A}} J_V(Y)$ we mean $\lim_{\Lambda \uparrow S}[\sum_{A \subset X \subset \Lambda} J_V(X)]$, with the implication that the limit exists. Let $\Lambda \in \mathscr{C}(S)$ with $\Lambda \supset A$. Then we have

$$\sum_{B \subset A} (-1)^{|B|} \sum_{B \subset Y \subset \Lambda} J_V(Y) = \sum_{Y \subset \Lambda} J_V(Y) \sum_{B \subset Y \cap A} (-1)^{|B|} = \sum_{Y \subset \Lambda - A} J_V(Y).$$

Therefore $\sum_{B \subset \Lambda - A} J_V(B) - \sum_{B \subset \Lambda} J_V(B)$

$$= \sum_{B \subset A} (-1)^{|B|} \sum_{B \subset Y \subset \Lambda} J_V(Y) - \sum_{X \subset \Lambda} J_V(Y)$$

$$= \sum_{\varnothing \neq B \subset A} (-1)^{|B|} \sum_{B \subset Y \subset \Lambda} J_V(Y),$$

and thus

$$\begin{aligned} \overline{V}(A) &= V(S-A) - V(S) = \lim_{\Lambda \uparrow S} (V(\Lambda - A) - V(\Lambda)) \\ &= \lim_{\Lambda \uparrow S} (\sum_{B \subset \Lambda - A} J_V(B) - \sum_{B \subset \Lambda} J_V(B)) \\ &= \lim_{\Lambda \uparrow S} \sum_{\varnothing \neq B \subset A} (-1)^{|B|} \sum_{B \subset Y \subset \Lambda} J_V(Y). \end{aligned}$$

It is now easily seen that

$$\lim_{\Lambda \uparrow S} (\sum_{B \subset Y \subset \Lambda} J_V(Y))$$

exists if $B \neq \varnothing$, and also that

$$\overline{V}(A) = \sum_{\varnothing \neq B \subset A} (-1)^{|B|} \lim_{\Lambda \uparrow S} \sum_{B \subset Y \subset \Lambda} J_V(Y).$$

Therefore we must have

$$J_V(A) = (-1)^{|A|} \lim_{\Lambda \uparrow S} \sum_{B \subset Y \subset \Lambda} J_V(Y) \quad \text{if} \quad A \neq \varnothing. \square$$

Combining Theorem 6.2 and Proposition 7.3 immediately gives:

PROPOSITION 7.4 Let $V \in H(S)$ and suppose there exists $\alpha < 1$ such that

$$\sum_{\substack{Y \in \mathscr{C}(S)\\ x \notin Y}} J_V(Y \cup x) - \sum_{\substack{\varnothing \neq A \in \mathscr{C}(S)\\ x \notin A}} \Big| \sum_{\substack{Y \supset \Lambda \cup x\\ Y \notin \mathscr{C}(S)}} J_V(Y) \Big| \geqslant \log\left(\frac{2}{\alpha}\right)$$

for all $x \in S$. Then phase transition cannot occur for V.

Proof It is easily checked that $\bar{V}$ satisfies the hypotheses of Theorem 6.2 and thus phase transition does not occur for $\bar{V}$. Therefore phase transition cannot occur for V.□

Let us for the moment consider the somewhat simple case of pair potentials. We say that $V \in H(S)$ is a *pair potential* if $J_V(A) = 0$ whenever $|A| > 2$. If V is a pair potential then we can define a symmetric bilinear form $U: S \times S \to \mathbb{R}$ by

$$U(x,y) = \begin{cases} \frac{1}{2} J_V(\{x,y\}) & \text{if} \quad x \neq y, \\ J_V(\{x\}) & \text{if} \quad x = y. \end{cases}$$

Then for any $A \in \mathscr{C}(S)$ we have $V(A) = U(A,A)$, where we use the notation that for any $A, B \in \mathscr{C}(S)$

$$U(A,B) = \sum_{x \in A} \sum_{y \in B} U(x,y).$$

Conversely if $U: S \times S \to \mathbb{R}$ is a symmetric bilinear form then defining $V(A) = U(A,A)$ gives us a pair potential V We will call U the bilinear form associated with V. Note from Proposition 7.3 that if V is a pair potential then $\bar{V}$ is a pair potential. Also if U (resp. $\bar{U}$) is the bilinear form associated with V (resp. $\bar{V}$) then

$$\bar{U}(x,y) = \begin{cases} U(x,y) & \text{if} \quad x \neq y, \\ U(x,x) - 2U(x,S) & \text{if} \quad x = y. \end{cases}$$

Hence for $A \in \mathscr{C}(S)$ we have

$$\bar{U}(A,A) = U(A,A) - 2U(A,S).$$

For pair potentials, Theorem 6.2 and Proposition 7.4 become:

Proposition 7.5. Let $V \in H(S)$ be a pair potential and let U be the bilinear form associated with V; let $\alpha < 1$. Then phase transition cannot occur for V if either:

(i) $U(x,x) + 2 \sum_{y \neq x} |U(x,y)| \leqslant \log\left(\frac{\alpha}{2}\right)$ for all $x \in S$,

or

(ii) $-U(x,x) + 2 \sum_{y \neq x} (|U(x,y)| - U(x,y)) \leqslant \log\left(\frac{\alpha}{2}\right)$ for all $x \in S$.

Proof (i) is just Theorem 6.2 and (ii) is Theorem 6.2 applied to $\bar{V}$.□

We will call $V \in H(S)$ an *Ising potential* if $\bar{V} = V$. If V is a pair potential with associated bilinear form U then clearly V is an Ising potential if and only if

$$U(x, S) = 0 \quad \text{for all} \quad x \in S.$$

More generally by Proposition 7.3 we have that $V \in H(S)$ is an Ising potential if and only if for all $A \in \mathscr{C}(S)$ with $A \neq \varnothing$ we have

$$J_V(A) = (-1)^{|A|} \sum_{\substack{Y \in \mathscr{C}(S) \\ Y \supset A}} J_V(Y).$$

For some applications it will be convenient to express a potential $V \in H(S)$ in a different way than we have done so far. To do this we need a few more definitions. If $A \in \mathscr{C}(S)$ then we define $\sigma_A : \mathscr{C}(S) \to \{-1, 1\}$ by

$$\sigma_A(B) = (-1)^{|A \cap B|} \quad \text{for} \quad B \in \mathscr{C}(S).$$

Let $A \subset \Lambda \in \mathscr{C}(S)$; then it is easy to check that

$$\sum_{B \subset \Lambda} \sigma_A(B) = \begin{cases} 2^{|\Lambda|} & \text{if} \quad A \neq \varnothing, \\ 0 & \text{otherwise.} \end{cases}$$

In order to avoid any problems about convergence we will assume from now on that we have a potential $V \in H(S)$ such that

$$\sum_{\substack{Y \in \mathscr{C}(S) \\ Y \supset A}} |J_V(Y)| \, 2^{-|Y|} < \infty \quad \text{for all} \quad A \in \mathscr{C}(S) \quad \text{with} \quad A \neq \varnothing.$$

We leave it to the reader to check whether the assumption is needed in what follows. Note that the assumption implies that

$$\lim_{\Lambda \to S} \sum_{\varnothing \neq X \in \mathscr{C}(S - \Lambda)} |J_V(X \cup A)| \, 2^{-|X|} = 0 \quad \text{for all} \quad A \in \mathscr{C}(S), A \neq \varnothing.$$

We now define $I_V : \mathscr{C}(S) \to \mathbb{R}$ by

$$I_V(E) = (-1)^{|E|} \sum_{\substack{A \in \mathscr{C}(S) \\ A \supset E}} 2^{-|A|} J_V(A) \quad \text{for} \quad E \neq \varnothing,$$

$$I_V(\varnothing) = 0.$$

PROPOSITION 7.6 Let $A \subset \Lambda \in \mathscr{C}(S)$. Then

$$\sum_{B \subset \Lambda} [\sigma_A(B) - \sigma_\varnothing(B)] \, I_V(B) = V(A) + \sum_{\substack{\varnothing \neq E \cap \Lambda \subset A \\ E \cap (S - \Lambda) \neq \varnothing}} 2^{-|E - A|} J_V(E);$$

and thus in particular

$$V(A) = \sum_{B\in\mathscr{C}(S)} [\sigma_A(B)-\sigma_\varnothing(B)]\, I_V(B).$$

Proof We have

$$\sum_{B\subset\Lambda} \sigma_A(B)\, I_V(B) = \sum_{\varnothing\neq B\subset\Lambda} \sigma_A(B)\,(-1)^{|B|} \sum_{\substack{E\in\mathscr{C}(S)\\ E\supset B}} 2^{-|E|} J_V(E)$$

$$= \sum_{\substack{E\in\mathscr{C}(S)\\ E\cap\Lambda\neq\varnothing}} J_V(E)\, 2^{-|E|} \sum_{\varnothing\neq B\subset E\cap\Lambda} \sigma_A(B)\,(-1)^{|B|}$$

$$= \sum_{\substack{E\in\mathscr{C}(S)\\ E\cap\Lambda\neq\varnothing}} J_V(E) 2^{-|E|} \sum_{B\subset E\cap\Lambda} \sigma_A(B)\,(-1)^{|B|} - \sum_{\substack{E\in\mathscr{C}(S)\\ E\cap\Lambda\neq\varnothing}} J_V(E) 2^{-|E|}.$$

Now

$$\sum_{B\subset E\cap\Lambda} \sigma_A(B)\,(-1)^{|B|} = \sum_{B\subset\Lambda\cap E} (-1)^{|B\cap A|}\,(-1)^{|B|}$$

$$= \sum_{B\subset\Lambda\cap E} (-1)^{|B\cap(A\cap E)|}\,(-1)^{|B|} = \sum_{B\subset\Lambda\cap E} (-1)^{|B\cap(\Lambda-A)\cap E|}$$

$$= \sum_{B\subset\Lambda\cap E} \sigma_B((\Lambda-A)\cap E)$$

$$= \begin{cases} 2^{|A\cap E|} & \text{if } E\cap\Lambda\subset A, \\ 0 & \text{otherwise.} \end{cases}$$

Therefore

$$\sum_{B\subset\Lambda} \sigma_A(B)\, I_V(B) = \sum_{\substack{E\in\mathscr{C}(S)\\ \varnothing\neq E\cap\Lambda\subset A}} J_V(E)\, 2^{|\Lambda\cap E|-|E|} - \sum_{\substack{E\in\mathscr{C}(S)\\ E\cap\Lambda\neq\varnothing}} J_V(E)\, 2^{-|E|}$$

$$= \sum_{E\subset A} J_V(E) + \sum_{\substack{\varnothing\neq E\cap\Lambda\subset A\\ E\cap(S-\Lambda)\neq\varnothing}} J_V(E)\, 2^{-|E-\Lambda|} - \sum_{\substack{E\in\mathscr{C}(S)\\ E\cap\Lambda\neq\varnothing}} J_V(E) 2^{-|E|}$$

$$= V(A) + \sum_{\substack{\varnothing\neq E\cap\Lambda\subset A\\ E\cap(S-\Lambda)\neq\varnothing}} J_V(E)\, 2^{-|E-\Lambda|} - \sum_{\substack{E\in\mathscr{C}(S)\\ E\cap\Lambda\neq\varnothing}} J_V(E)\, 2^{-|E|}.$$

In particular we have

$$\sum_{B\subset\Lambda} \sigma_\varnothing(B)\, I_V(B) = -\sum_{\substack{E\in\mathscr{C}(S)\\ E\cap\Lambda\neq\varnothing}} J_V(E)\, 2^{-|E|},$$

and thus the result follows. □

It is not difficult to compute J_V in terms of I_V.

PROPOSITION 7.7 If $E\in\mathscr{C}(S)$ with $E\neq\varnothing$ then

$$J_V(E) = (-1)^{|E|} 2^{|E|} \sum_{\substack{B\in\mathscr{C}(S)\\ B\supset E}} I_V(B).$$

Proof Let $\varnothing \neq E \subset \Lambda \in \mathscr{C}(S)$; then

$$\sum_{E \subset B \subset \Lambda} I_V(B) = \sum_{E \subset B \subset \Lambda} (-1)^{|B|} \sum_{\substack{A \in \mathscr{C}(S) \\ A \supset B}} 2^{-|A|} J_V(A)$$

$$= \sum_{\substack{A \in \mathscr{C}(S) \\ A \supset E}} 2^{-|A|} J_V(A) \sum_{E \subset B \subset A \cap \Lambda} (-1)^{|B|}$$

$$= \sum_{\substack{A \in \mathscr{C}(S) \\ A \cap \Lambda = E}} (-1)^{|E|} 2^{-|A|} J_V(A)$$

$$= (-1)^{|E|} 2^{-|E|} J_V(E) + (-1)^{|E|} \sum_{\substack{A \cap \Lambda = E \\ A \cap (S-\Lambda) \neq \varnothing}} 2^{-|A|} J_V(A).$$

Thus we have

$$\lim_{\Lambda \uparrow S} \sum_{E \subset B \subset \Lambda} I_V(B) = (-1)^{|E|} 2^{-|E|} J_V(E). \square$$

Note that by Proposition 7.7 and from the definition of I_V we have that if $n \geqslant 1$ then $J_V(A) = 0$ for all $A \in \mathscr{C}(S)$ with $|A| \geqslant n$ if and only if $I_V(A) = 0$ for all $A \in \mathscr{C}(S)$ with $|A| \geqslant n$. In particular if V is a pair potential then $I_V(A) = 0$ for all $A \in \mathscr{C}(S)$ with $|A| > 2$.

PROPOSITION 7.8 We have

(i) $I_{\overline{V}}(E) = \sum_{A \supset E} 2^{-|A|} J_V(A)$ for $E \in \mathscr{C}(S)$ with $E \neq \varnothing$.

(ii) $I_{\overline{V}}(E) = (-1)^{|E|} I_V(E)$ for all $E \in \mathscr{C}(S)$.

(iii) V is an Ising potential if and only if $I_V(E) = 0$ for all $E \in \mathscr{C}(S)$ with $|E|$ odd.

Proof It is clear that (i) $\Rightarrow$ (ii) $\Rightarrow$ (iii). If $E \in \mathscr{C}(S)$ with $E \neq \varnothing$ then

$$I_{\overline{V}}(E) = (-1)^{|E|} \sum_{\substack{A \in \mathscr{C}(S) \\ A \supset E}} 2^{-|A|} J_{\overline{V}}(A)$$

$$= (-1)^{|E|} \sum_{\substack{A \in \mathscr{C}(S) \\ A \supset E}} 2^{-|A|} (-1)^{|A|} \sum_{\substack{B \in \mathscr{C}(S) \\ B \supset A}} J_V(B)$$

$$= (-1)^{|E|} \sum_{\substack{B \in \mathscr{C}(S) \\ B \supset E}} J_V(B) \sum_{E \subset A \subset B} (-1)^{|A|} 2^{-|A|}$$

$$= (-1)^{|E|} \sum_{\substack{B \in \mathscr{C}(S) \\ B \supset E}} J_V(B) (-1)^{|E|} 2^{-|B|} = \sum_{\substack{B \in \mathscr{C}(S) \\ B \supset E}} 2^{-|B|} J_V(B). \square$$

8. *Attractive and supermodular potentials*

In this chapter we look at potentials having some additional properties, and for these potentials the problem of when phase transition occurs becomes easier to handle.

A potential $V \in H(S)$ will be called *supermodular* if for all $A, B \in \mathscr{C}(S)$ we have

$$V(A \cup B) + V(A \cap B) \geqslant V(A) + V(B).$$

V will be called *attractive* if $J_V(A) \geqslant 0$ for all $A \in \mathscr{C}(S)$ with $|A| \geqslant 2$. (The term supermodular comes from the defining property; the term attractive comes from the type of interaction a model with this kind of potential would describe.) We first note that the property of being attractive is stronger than that of being supermodular.

PROPOSITION 8.1 If $V \in H(S)$ is attractive then it is supermodular.

Proof Let $A, B \in \mathscr{C}(S)$. Then

$$\begin{aligned} & V(A \cup B) + V(A \cap B) - V(A) - V(B) \\ &= \sum_{X \subset A \cup B} J_V(X) + \sum_{X \subset A \cap B} J_V(X) - \sum_{X \subset A} J_V(X) - \sum_{X \subset B} J_V(X) \\ &= \sum_{X \in \Omega} J_V(X), \end{aligned}$$

where $\Omega = \{X \subset A \cup B : X \cap (A - B) \neq \varnothing, X \cap (B - A) \neq \varnothing\}$. If $X \in \Omega$ then $|X| \geqslant 2$ and thus $J_V(X) \geqslant 0$. Therefore

$$V(A \cup B) + V(A \cap B) \geqslant V(A) + V(B). \square$$

Although in general the converse of Proposition 8.1 is clearly false, it is true for pair potentials.

PROPOSITION 8.2 Let $V \in H(S)$ be a pair potential. Then V is attractive if and only if it is supermodular.

Proof Let V be supermodular and take $x, y \in S$ with $x \neq y$. Then

$$J_V(\{x,y\}) = V(\{x\} \cup \{y\}) - V(\{x\}) - V(\{y\}) + V(\{x\} \cap \{y\}) \geqslant 0,$$

and thus V is attractive. The converse has of course already been proved. □

The most important tool we have for examining supermodular potentials is Holley's inequality (Theorem 3.2) which we recall is: If Λ is a finite set and μ_1, μ_2 are strictly positive probability densities on $\mathscr{P}(\Lambda)$ (i.e. for $i = 1, 2$ we have $\mu_i: \mathscr{P}(\Lambda) \to \mathbb{R}$ with $\mu_i(A) > 0$ for all $A \in \mathscr{P}(\Lambda)$ and

$$\sum_{A \in \mathscr{P}(\Lambda)} \mu_i(A) = 1)$$

and if for all $A, B \in \mathscr{P}(\Lambda)$ we have

$$\mu_1(A \cup B)\, \mu_2(A \cap B) \geqslant \mu_1(A)\, \mu_2(B),$$

then for any increasing $f: \mathscr{P}(\Lambda) \to \mathbb{R}$ (i.e. if $A \supset B$ then $f(A) \geqslant f(B)$) we have

$$\sum_{A \subset \Lambda} f(A)\, \mu_1(A) \geqslant \sum_{A \subset \Lambda} f(A)\, \mu_2(A).$$

Let $\mathscr{V} \in D_+(S)$ with $\mathscr{V} = \{f^\Lambda\}_{\Lambda \in \mathscr{C}(S)}$ and let $V \in H(S)$ be the potential associated with $\mathscr{V}$. We recall the following definitions from Chapter 5. If $\Lambda \in \mathscr{C}(S)$ and $X \in \mathscr{P}(S - \Lambda)$ then $\pi_\Lambda^X \in \mathscr{S}(\Lambda)$ was given by

$$\pi_\Lambda^X([A, \Lambda]) = f^\Lambda(A, X) \quad \text{for} \quad A \subset \Lambda.$$

The correlation function of π_Λ^X was denoted by ρ_Λ^X, i.e.

$$\rho_\Lambda^X : \mathscr{C}(\Lambda) \to \mathbb{R}$$

is defined by

$$\rho_\Lambda^X(A) = \sum_{A \subset B \subset \Lambda} f^\Lambda(B, X) \quad \text{for} \quad A \subset \Lambda.$$

We also defined $\rho_\Lambda^+ : \mathscr{C}(\Lambda) \to \mathbb{R}$, $\rho_\Lambda^- : \mathscr{C}(\Lambda) \to \mathbb{R}$ by

$$\rho_\Lambda^+(A) = \max_{X \in \mathscr{P}(S-\Lambda)} \rho_\Lambda^X(A),$$

$$\rho_\Lambda^-(A) = \min_{X \in \mathscr{P}(S-\Lambda)} \rho_\Lambda^X(A);$$

and we noted that in general ρ_Λ^+ and ρ_Λ^- are not correlation functions. However if the potential V is supermodular then ρ_Λ^+ and ρ_Λ^- are correlation functions, since we have:

PROPOSITION 8.3 Let V be supermodular and $A \subset \Lambda \in \mathscr{C}(S)$. Then $\rho_\Lambda^X(A)$ is an increasing function of X; i.e. if $X, Y \in \mathscr{P}(S-\Lambda)$ with $X \subset Y$ then $\rho_\Lambda^X(A) \leqslant \rho_\Lambda^Y(A)$.

Proof Since π_Λ^X is a measure on a finite set we can identify it with its density, i.e. we will write $\pi_\Lambda^X(B)$ instead of $\pi_\Lambda^X([B,\Lambda])$. By continuity we can assume that $X, Y \in \mathscr{C}(S-\Lambda)$. If $E, F \in \mathscr{C}(\Lambda)$ then since V is supermodular we have that

$$V(E \cup F \cup Y) + V((E \cap F) \cup X) = V((E \cup Y) \cup (F \cup X)) + V((E \cup Y) \cap (F \cup X)) \geqslant V(E \cup Y) + V(F \cup X),$$

and thus

$$\exp V(E \cup F \cup Y) \exp V((E \cap F) \cup X) \geqslant \exp V(E \cup Y) \exp V(F \cup X).$$

Therefore
$$\pi_\Lambda^Y(E \cup F)\, \pi_\Lambda^X(E \cap F) \geqslant \pi_\Lambda^Y(E)\, \pi_\Lambda^X(F),$$

i.e. π_Λ^Y and π_Λ^X satisfy the hypotheses of Holley's inequality. Now define $f: \mathscr{P}(\Lambda) \to \mathbb{R}$ by

$$f(B) = \begin{cases} 1 & \text{if } A \subset B \subset \Lambda, \\ 0 & \text{otherwise.} \end{cases}$$

Clearly f is increasing and we have

$$\rho_\Lambda^X(A) = \sum_{B \subset \Lambda} f(B)\, \pi_\Lambda^X(B),$$

$$\rho_\Lambda^Y(A) = \sum_{B \subset \Lambda} f(B)\, \pi_\Lambda^Y(B).$$

Therefore by Holley's inequality we have $\rho_\Lambda^Y(A) \geqslant \rho_\Lambda^X(A)$. □

It immediately follows from Proposition 8.3 that

$$\rho_\Lambda^+(A) = \rho_\Lambda^{S-\Lambda}(A) \quad \text{and} \quad \rho_\Lambda^-(A) = \rho_\Lambda^{\varnothing}(A).$$

Thus ρ_Λ^+ is the correlation function $\rho_\Lambda^{S-\Lambda}$ and ρ_Λ^- is the correlation function $\rho_\Lambda^\varnothing$. Recall that Proposition 5.6 states that $\rho_\Lambda^+(A)$ is a decreasing function of Λ and $\rho_\Lambda^-(A)$ is an increasing function of Λ; thus we could define $\rho^+ : \mathscr{C}(S) \to \mathbb{R}$, $\rho^- : \mathscr{C}(S) \to \mathbb{R}$ by

$$\rho^+(A) = \lim_{\Lambda \uparrow S} \rho_\Lambda^+(A),$$

$$\rho^-(A) = \lim_{\Lambda \uparrow S} \rho_\Lambda^-(A).$$

For supermodular potentials V we actually have that ρ^+ and ρ^- are correlation functions of Gibbs states.

PROPOSITION 8.4 If V is supermodular then there exist $\mu^+, \mu^- \in \mathscr{G}_V$ such that ρ^+ (resp. ρ^-) is the correlation function of μ^+ (resp. μ^-). Furthermore, if for any $\Lambda \in \mathscr{C}(S)$ we define

$$\mu_\Lambda^+, \mu_\Lambda^- \in \mathscr{S}(S)$$

by
$$\mu_\Lambda^+([A, \Lambda]) = \pi_\Lambda^{S-\Lambda}(A) \quad \text{for} \quad A \subset \Lambda,$$
$$\mu_\Lambda^+[(x, x]) = 1 \quad \text{for} \quad x \in S - \Lambda;$$
and
$$\mu_\Lambda^-([A, \Lambda]) = \pi_\Lambda^\varnothing(A) \quad \text{for} \quad A \subset \Lambda,$$
$$\mu_\Lambda^-([\varnothing, x]) = 1 \quad \text{for} \quad x \in S - \Lambda;$$
then
$$\mu_\Lambda^+ \to \mu^+, \mu_\Lambda^- \to \mu^- \quad \text{(in the vague topology)}$$
as $\Lambda \uparrow S$.

Proof For any $A \subset B \in \mathscr{C}(S)$ it is clear that $\mu_\Lambda^+([A, B])$ converges as $\Lambda \uparrow S$. Thus there exists μ^+ such that $\mu_\Lambda^+ \to \mu^+$ as $\Lambda \uparrow S$, and it is easy to check that ρ^+ must be the correlation function of μ^+. Now by Lemma 5.1 we have $\mu^+ \in \mathscr{G}_V$. Exactly the same proof works of course for ρ^-. □

If V is supermodular then we will call μ^+ (resp. μ^-) as given in Proposition 8.4 the *high* (resp. *low*) *density Gibbs state* with potential V. By Theorem 5.1 we have that phase transition occurs for V if and only if $\mu^+ \neq \mu^-$. We can make a further simplication to the phase transition problem for supermodular potentials by showing that $\rho^+ - \rho^-$ is subadditive.

Proposition 8.5 If V is supermodular then for all $A, B \in \mathscr{C}(S)$ we have

$$(\rho^+(A \cup B) - \rho^-(A \cup B)) \leqslant (\rho^+(A) - \rho^-(A)) + (\rho^+(B) - \rho^-(B)).$$

Proof Let $A, B \in \mathscr{C}(S)$ and take $\Lambda \in \mathscr{C}(S)$ with $\Lambda \supset A \cup B$. Then

$$\rho^+_\Lambda(A) + \rho^+_\Lambda(B) - \rho^+_\Lambda(A \cup B) = \sum_{E \in \Omega} \pi^{S-\Lambda}_\Lambda(E),$$

where $\Omega = \{E \subset \Lambda\colon \text{either } E \supset A \text{ or } E \supset B\}$. Define $f\colon \mathscr{P}(\Lambda) \to \mathbb{R}$ by

$$f(E) = \begin{cases} 1 & \text{if} \quad E \supset A \quad \text{or} \quad E \supset B, \\ 0 & \text{otherwise.} \end{cases}$$

Clearly f is increasing and we have

$$\rho^+_\Lambda(A) + \rho^+_\Lambda(B) - \rho^+_\Lambda(A \cup B) = \sum_{E \subset \Lambda} f(E)\, \pi^{S-\Lambda}_\Lambda(E).$$

Similarly we also have

$$\rho^-_\Lambda(A) + \rho^-_\Lambda(B) - \rho^-_\Lambda(A \cup B) = \sum_{E \subset \Lambda} f(E)\, \pi^{\varnothing}_\Lambda(E).$$

But by the proof of Proposition 8.3 we have that $\pi^{S-\Lambda}_\Lambda$ and $\pi^{\varnothing}_\Lambda$ satisfy the hypotheses of Holley's inequality, and thus by Holley's inequality we have

$$\sum_{E \subset \Lambda} f(E)\, \pi^{S-\Lambda}_\Lambda(E) \geqslant \sum_{E \subset \Lambda} f(E)\, \pi^{\varnothing}_\Lambda(E).$$

Therefore

$$\rho^+_\Lambda(A) + \rho^+_\Lambda(B) - \rho^+_\Lambda(A \cup B) \geqslant \rho^-_\Lambda(A) + \rho^-_\Lambda(B) - \rho^-_\Lambda(A \cup B),$$

and hence the result follows on letting $\Lambda \uparrow S$. □

An immediate application of the above proposition is that for all $A \in \mathscr{C}(S)$ we have

$$\begin{aligned} \rho^+(A) - \rho^-(A) &\leqslant \sum_{x \in A} [\rho^+(\{x\}) - \rho^-(\{x\})] \\ &\leqslant |A| \max_{x \in A} [\rho^+(\{x\}) - \rho^-(\{x\})]. \end{aligned}$$

Therefore $\rho^+ = \rho^-$ if and only if $\rho^+(\{x\}) = \rho^-(\{x\})$ for all $x \in S$; and thus we have proved:

THEOREM 8.1 If V is supermodular then phase transition occurs for V if and only if $\rho^+(\{x\}) \neq \rho^-(\{x\})$ for some $x \in S$.

We now look again at the Kirkwood–Salsburg equations. From the proof of Theorem 6.2 it is clear that things would be a lot simpler if given $x \in \Lambda \in \mathscr{C}(S)$ and $B \in \mathscr{C}(S-\Lambda)$ then

$$\sum_{X \subset B} (-1)^{|B-X|} \exp [V(\Lambda \cup X) - V((\Lambda - x) \cup X)] \geqslant 0.$$

A sufficient condition for this to happen is given by:

PROPOSITION 8.6 Suppose that V is attractive. Then given $x \in \Lambda \in \mathscr{C}(S)$ and $B \in \mathscr{C}(S-\Lambda)$ we have

$$\sum_{X \subset B} (-1)^{|B-X|} \exp [V(\Lambda \cup X) - V((\Lambda - x) \cup X)] \geqslant 0.$$

Proof Just as in the proof of Theorem 6.2 we have

$$V(\Lambda \cup X) - V((\Lambda - x) \cup X) = \sum_{Y \subset X} \sum_{E \subset (\Lambda - x)} J_V(E \cup x \cup Y)$$

and by hypotheses if $Y \neq \varnothing$ then

$$\sum_{E \subset (\Lambda - x)} J_V(E \cup x \cup Y) \geqslant 0.$$

The result now follows from Lemma 6.2 exactly as in the proof of Theorem 6.2. □

Using this proposition we can improve on Theorem 6.2 when the potential V is attractive.

PROPOSITION 8.7 Let V be attractive and let

$$\alpha = \sup_{A \in \mathscr{C}(S)} (\rho^+(A) - \rho^-(A)).$$

Then given $x \in \Lambda \in \mathscr{C}(S)$ we have

$$\rho^+(\Lambda) - \rho^-(\Lambda) \leqslant \alpha[1 + \exp(V(\Lambda) - V(\Lambda - x))]^{-1} \exp(-\overline{V}(\{x\})).$$

Proof By Theorem 6.1 we have

$$\rho^+(\Lambda) - \rho^-(\Lambda) = \lim_{\Lambda' \to S} \sum_{B \subset \Lambda' - \Lambda} [\sum_{X \subset B} (-1)^{|B-X|} \exp(V(\Lambda \cup X)$$
$$- V((\Lambda - x) \cup X))] (\rho^+((\Lambda - x) \cup B) - \rho^-((\Lambda - x) \cup B)$$
$$- \rho^+(\Lambda \cup B) + \rho^-(\Lambda \cup B))$$

$$\leqslant \lim_{\Lambda' \to S} \sum_{B \subset \Lambda' - \Lambda} \Big[\sum_{X \subset B} (-1)^{|B-X|} \exp\left(V(\Lambda \cup X) - V((\Lambda - x) \cup X)\right)\Big]$$
$$\times (\rho^+((\Lambda - x) \cup B) - \rho^-((\Lambda - x) \cup B))$$
$$- \exp\left(V(\Lambda) - V(\Lambda - x)\right) (\rho^+(\Lambda) - \rho^-(\Lambda))$$
$$\leqslant \alpha \lim_{\Lambda' \to S} \sum_{B \subset \Lambda' - \Lambda} \Big[\sum_{X \subset B} (-1)^{|B-X|} \exp\left(V(\Lambda \cup X)\right.$$
$$\left.- V((\Lambda - x) \cup X)\right)\Big] - (\rho^+(\Lambda) - \rho^-(\Lambda)) \exp\left(V(\Lambda) - V(\Lambda - x)\right)$$
$$= \alpha \lim_{\Lambda' \to S} \exp\left(V(\Lambda') - V(\Lambda' - x)\right)$$
$$- (\rho^+(\Lambda) - \rho^-(\Lambda)) \exp\left(V(\Lambda) - V(\Lambda - x)\right).$$

But
$$\lim_{\Lambda' \to S} (V(\Lambda') - V(\Lambda' - x)) = \overline{V}(\{x\}),$$
and thus

$$(\rho^+(\Lambda) - \rho^-(\Lambda)) (1 + \exp\left(V(\Lambda) - V(\Lambda - x)\right)) \leqslant \alpha \exp\left(-\overline{V}(\{x\})\right). \square$$

THEOREM 8.2 If V is attractive and
$$\inf_{x \in S} [\exp \overline{V}(\{x\}) (1 + \exp V(\{x\}))] > 1$$
then phase transition cannot occur for V.

Proof Let
$$\alpha = \sup_{A \in \mathscr{C}(S)} (\rho^+(A) - \rho^-(A)).$$
If $x \in \Lambda \in \mathscr{C}(S)$ then since V is attractive we have
$$V(\Lambda) - V(\Lambda - x) \geqslant V(\{x\}),$$
and thus by Proposition 8.7
$$\rho^+(\Lambda) - \rho^-(\Lambda) \leqslant \alpha[1 + \exp V(\{x\})]^{-1} \exp\left(-\overline{V}(\{x\})\right).$$
Therefore $\rho^+ = \rho^-$ if
$$\sup_{x \in S} [1 + \exp V(\{x\})]^{-1} \exp\left(-\overline{V}(\{x\})\right) < 1,$$
i.e. if
$$\inf_{x \in S} [\exp \overline{V}(\{x\}) (1 + \exp V(\{x\}))] > 1. \square$$

Note that if V is an attractive Ising potential then (since $V(\{x\}) = \overline{V}(\{x\})$) we have that phase transition cannot occur for V if
$$\inf_{x \in S} V(\{x\}) > \log\left(\frac{\sqrt{5} - 1}{2}\right).$$

We will now consider some more results for supermodular potentials. If V is supermodular then it is easy to check that $\bar{V}$ is also supermodular (although note that this would not be true in general if supermodular was replaced by attractive). Let $\pi_\Lambda^X, \rho_\Lambda^X, \rho_\Lambda^+, \rho_\Lambda^-, \rho^+, \rho^-$ be defined as before in terms of V and let $\bar{\pi}_\Lambda^X, \bar{\rho}_\Lambda^X, \bar{\rho}_\Lambda^+, \bar{\rho}_\Lambda^-, \bar{\rho}^+, \bar{\rho}^-$ be the corresponding expressions defined in terms of $\bar{V}$. The following facts, which we list as a proposition, are easily verified.

PROPOSITION 8.8 (i) If $A \subset \Lambda \in \mathscr{C}(S)$ and $X \in \mathscr{P}(S-\Lambda)$ then

$$\pi_\Lambda^X(A) = \bar{\pi}_\Lambda^{S-X}(\Lambda - A).$$

(ii) If V is supermodular and $x \in \Lambda \in \mathscr{C}(S)$ then

$$\bar{\rho}_\Lambda^+(\{x\}) = 1 - \rho_\Lambda^-(\{x\}).$$

(iii) If V is supermodular and $x \in S$ then $\bar{\rho}^+(\{x\}) = 1 - \rho^-(\{x\})$.

Proof (i) follows from the definition of $\bar{V}$; (ii) clearly implies (iii), and (ii) is true since

$$\bar{\rho}_\Lambda^+(\{x\}) = \sum_{x \in A \subset \Lambda} \bar{\pi}_\Lambda^{S-\Lambda}(A) = \sum_{x \in A \subset \Lambda} \pi_\Lambda^\varnothing(\Lambda - A)$$

$$= \sum_{x \notin B \subset \Lambda} \pi_\Lambda^\varnothing(B) = 1 - \sum_{x \in B \subset \Lambda} \pi_\Lambda^\varnothing(B) = 1 - \rho_\Lambda^-(\{x\}). \square$$

On applying the above proposition to Ising potentials we get:

PROPOSITION 8.9 Let V be a supermodular Ising potential. Then for any $x \in S$ we have

$$\rho^+(\{x\}) = 1 - \rho^-(\{x\}).$$

Therefore as $\rho^-(\{x\}) \leqslant \rho^+(\{x\})$ we must have

$$\rho^-(\{x\}) \leqslant \tfrac{1}{2} \leqslant \rho^+(\{x\}),$$

and phase transition occurs for V if and only if $\rho^-(\{x\}) < \frac{1}{2}$ for some $x \in S$.

Proof This is clear. $\square$

Let now V be a potential such that

$$\sum_{\substack{Y \in \mathscr{C}(S) \\ Y \supset A}} |J_V(Y)|\, 2^{-|Y|} < \infty \quad \text{for all} \quad A \in \mathscr{C}(S) \quad \text{with} \quad A \neq \varnothing.$$

Thus as in Chapter 7 we can define $I_V: \mathscr{C}(S) \to \mathbb{R}$ by

$$I_V(E) = (-1)^{|E|} \sum_{\substack{A \in \mathscr{C}(S) \\ A \supset E}} 2^{-|A|} J_V(A) \quad \text{for} \quad E \neq \varnothing,$$

$$I_V(\varnothing) = 0.$$

To examine potentials in terms of the function I_V it will be useful to have some form of the Griffiths inequalities, which we will now describe. Firstly recall that if $A, B \in \mathscr{C}(S)$ then we let AB denote the symmetric difference of A and B, i.e.

$$AB = (A-B) \cup (B-A).$$

It is easy to check that this product makes $\mathscr{C}(S)$ a commutative group since $AB = BA$, $(AB)C = A(BC)$, $A\varnothing = A$ and $AA = \varnothing$. Thus $\varnothing$ is the identity in the group and each element is its own inverse. Note that for any $X \in \mathscr{P}(S)$ we have $\mathscr{C}(X)$ is a subgroup of $\mathscr{C}(S)$; in particular if $\Lambda \in \mathscr{C}(S)$ and $f: \mathscr{P}(\Lambda) \to \mathbb{R}$ then

$$\sum_{A \subset \Lambda} f(A) = \sum_{A \subset \Lambda} f(AE) \quad \text{for any} \quad E \subset \Lambda.$$

Recall also that for $A \in \mathscr{C}(S)$ we defined $\sigma_A: \mathscr{C}(S) \to \{-1, 1\}$ by

$$\sigma_A(B) = (-1)^{|A \cap B|} \quad \text{for} \quad B \in \mathscr{C}(S).$$

It is clear that $\sigma_A(B) = \sigma_B(A)$ and it is a simple matter to check that if $A, B, E \in \mathscr{C}(S)$ then

$$\sigma_A(E)\,\sigma_B(E) = \sigma_{AB}(E).$$

Finally if $A \in \mathscr{C}(S)$ and $\Lambda \in \mathscr{C}(S)$ then

$$\sum_{B \subset \Lambda} \sigma_A(B) = \begin{cases} 2^{|\Lambda|} & \text{if} \quad A \cap \Lambda = \varnothing, \\ 0 & \text{otherwise.} \end{cases}$$

Let $I: \mathscr{C}(S) \to \mathbb{R}$ with $I(\varnothing) = 0$; and let $\Lambda \subset \Lambda' \in \mathscr{C}(S)$. We define an element of $\mathscr{S}(\Lambda)$ by letting its density $\pi: \mathscr{P}(\Lambda) \to \mathbb{R}$ be given by

$$\pi(A) = Z^{-1} \exp\Big[\sum_{B \subset \Lambda'} \sigma_A(B)\, I(B)\Big] \quad \text{for} \quad A \subset \Lambda,$$

where of course $Z = \sum_{A \subset \Lambda} \exp\Big[\sum_{B \subset \Lambda'} \sigma_A(B)\, I(B)\Big]$.

Now we have:

THEOREM 8.3 Suppose that $I(A) \geqslant 0$ for all $A \in \mathscr{C}(S)$. Then

(i) $\sum_{A\subset\Lambda} \sigma_E(A)\pi(A) \geqslant 0$ for all $E\in\mathscr{C}(S)$.

(ii) $\sum_{A\subset\Lambda} \sigma_E(A)\sigma_F(A)\pi(A) - \sum_{A\subset\Lambda}\sigma_E(A)\pi(A)\sum_{B\subset\Lambda}\sigma_F(B)\pi(B) \geqslant 0$

for all $E, F\in\mathscr{C}(S)$.

((i) and (ii) are called the *Griffiths inequalities.*)

Proof (i) We have

$$Z\sum_{A\subset\Lambda}\sigma_E(A)\pi(A) = \sum_{A\subset\Lambda}\sigma_E(A)\exp[\sum_{B\subset\Lambda'}\sigma_A(B)I(B)]$$

$$= \sum_{n=0}^{\infty}\frac{1}{n!}\sum_{A\subset\Lambda}\sigma_E(A)[\sum_{B_1\subset\Lambda'}\cdots\sum_{B_n\subset\Lambda'}\sigma_A(B_1)\ldots\sigma_A(B_n) \times I(B_1)\ldots I(B_n)]$$

$$= \sum_{n=0}^{\infty}\frac{1}{n!}[\sum_{B_1\subset\Lambda'}\cdots\sum_{B_n\subset\Lambda'}I(B_1)\ldots I(B_n)\sum_{A\subset\Lambda}\sigma_E(A) \times \sigma_A(B_1)\ldots\sigma_A(B_n)]$$

$$= \sum_{n=0}^{\infty}\frac{1}{n!}[\sum_{B_1\subset\Lambda'}\cdots\sum_{B_n\subset\Lambda'}I(B_1)\ldots I(B_n)\sum_{A\subset\Lambda}\sigma_{EB_1\ldots B_n}(A)]$$

$$= \sum_{n=0}^{\infty}\frac{1}{n!}[\sum_{\substack{B_1\subset\Lambda' \\ (EB_1\ldots B_n)\cap\Lambda=\varnothing}}\cdots\sum_{B_n\subset\Lambda'}I(B_1)\ldots I(B_n)\,2^{|\Lambda|}] \geqslant 0.$$

Therefore $\sum_{A\subset\Lambda}\sigma_E(A)\pi(A)\geqslant 0$.

(ii) $$Z^2\{\sum_{A\subset\Lambda}\sigma_E(A)\sigma_F(A)\pi(A) - \sum_{A\subset\Lambda}\sigma_E(A)\pi(A)\sum_{B\subset\Lambda}\sigma_F(A)\pi(B)\}$$

$$= \sum_{A\subset\Lambda}\sum_{B\subset\Lambda}(\sigma_E(A)\sigma_F(A)-\sigma_E(A)\sigma_F(B))\exp[\sum_{X\subset\Lambda'}\sigma_A(X)I(X)] \times\exp[\sum_{Y\subset\Lambda'}\sigma_B(Y)I(Y)]$$

$$= \sum_{A\subset\Lambda}\sum_{B\subset\Lambda}(\sigma_E(A)\sigma_F(A)-\sigma_E(A)\sigma_F(B)) \times\exp[\sum_{X\subset\Lambda'}(\sigma_A(X)+\sigma_B(X))I(X)]$$

$$= \sum_{A\subset\Lambda}\sum_{B\subset\Lambda}(\sigma_E(A)\sigma_F(A)-\sigma_E(A)\sigma_F(AB)) \times\exp[\sum_{X\subset\Lambda'}(\sigma_A(X)+\sigma_{AB}(X))I(X)]$$

$$= \sum_{B\subset\Lambda}(1-\sigma_F(B))\sum_{A\subset\Lambda}\sigma_E(A)\sigma_F(A) \times\exp[\sum_{X\subset\Lambda'}\sigma_A(X)(1+\sigma_B(X))I(X)].$$

Now for fixed $B \subset \Lambda$ we have from (i) that

$$\sum_{A \subset \Lambda} \sigma_E(A)\, \sigma_F(A) \exp\Big[\sum_{X \subset \Lambda'} \sigma_A(X)\,(1+\sigma_B(X))\, I(X)\Big] \geqslant 0,$$

since $\sigma_E(A)\,\sigma_F(A) = \sigma_{EF}(A)$ and $(1+\sigma_B(X))\, I(X) \geqslant 0$ for all $X \in \mathscr{C}(S)$. Also of course $(1-\sigma_F(B)) \geqslant 0$, and thus the result follows.□

We will use the Griffiths inequalities to prove the following: let $I: \mathscr{C}(S) \to \mathbb{R}$, $I': \mathscr{C}(S) \to \mathbb{R}$ with $I(\varnothing) = I'(\varnothing) = 0$, let

$$\Lambda \subset \Lambda' \in \mathscr{C}(S),$$

let π be as before and let π' be the corresponding density defined in terms of I'.

PROPOSITION 8.10 If $I'(A) \geqslant I(A) \geqslant 0$ for all $A \in \mathscr{C}(S)$ then for any $E \in \mathscr{C}(S)$ we have

$$\sum_{A \subset \Lambda} \sigma_E(A)\, \pi(A) \leqslant \sum_{A \subset \Lambda} \sigma_E(A)\, \pi'(A).$$

Proof We need only consider I, I' such that there exists $F \subset \Lambda'$ with $I(B) = I'(B)$ if $B \neq F$ (since π and π' are determined by the values of I and I' on the finite set $\mathscr{P}(\Lambda')$). For $0 \leqslant t \leqslant 1$ let $I_t: \mathscr{C}(S) \to \mathbb{R}$ be given by $I_t(A) = (1-t)\, I(A) + tI'(A)$. Thus in fact

$$I_t(A) = \begin{cases} I(A) & \text{if} \quad A \neq F, \\ I(F) + (I'(F) - I(F))\, t & \text{if} \quad A = F. \end{cases}$$

Let π_t be the density corresponding to I_t, and let

$$h(t) = \sum_{A \subset \Lambda} \sigma_E(A)\, \pi_t(A).$$

Then it is not hard to compute that

$$\frac{\mathrm{d}}{\mathrm{d}t}(h(t)) = (I'(F) - I(F))\Big\{\sum_{A \subset \Lambda} \sigma_E(A)\, \sigma_F(A)\, \pi_t(A) - \sum_{A \subset \Lambda} \sigma_E(A)\, \pi_t(A) \sum_{B \subset \Lambda} \sigma_t(B)\, \pi_t(B)\Big\}$$

and thus by the Griffiths inequalities we have

$$\frac{\mathrm{d}}{\mathrm{d}t}(h(t)) \geqslant 0.$$

In particular we have $h(0) \leqslant h(1)$, i.e.

$$\sum_{A \subset \Lambda} \sigma_E(A)\,\pi(A) \leqslant \sum_{A \subset \Lambda} \sigma_E(A)\,\pi'(A). \square$$

Let $V, \tilde{V} \in H(S)$ with

$$\sum_{\substack{Y \in \mathscr{C}(S) \\ Y \supset A}} |J_V(Y)|\, 2^{-|Y|} < \infty \quad \text{for all} \quad A \in \mathscr{C}(S) \quad \text{with} \quad A \neq \varnothing,$$

$$\sum_{\substack{Y \in \mathscr{C}(S) \\ Y \supset A}} |J_{\tilde{V}}(Y)|\, 2^{-|Y|} < \infty \quad \text{for all} \quad A \in \mathscr{C}(S) \quad \text{with} \quad A \neq \varnothing.$$

Thus we can define I_V and $I_{\tilde{V}}$ as before; for notational convenience we will write I (resp. $\tilde{I}$) instead of I_V (resp. $I_{\tilde{V}}$).

Proposition 8.11 Suppose that $\tilde{I}(A) \geqslant I(A) \geqslant 0$ for all $A \in \mathscr{C}(S)$; let $x \in \Lambda \in \mathscr{C}(S)$. Then

$$\rho_\Lambda^\varnothing(\{x\}) \geqslant \tilde{\rho}_\Lambda^\varnothing(\{x\}).$$

Proof Let $\Lambda' \in \mathscr{C}(S)$ with $\Lambda' \supset \Lambda$, and let π, $\tilde{\pi}$ be defined as in Proposition 8.10. Note that if $A \subset \Lambda$ then

$$\pi(A) = Z^{-1} \exp\Big[\sum_{B \subset \Lambda'} (\sigma_A(B) - \sigma_\varnothing(B))\, I(B)\Big],$$

with
$$Z = \sum_{A \subset \Lambda} \exp\Big[\sum_{B \subset \Lambda'} (\sigma_A(B) - \sigma_\varnothing(B))\, I(B)\Big]$$

(since the term $\exp\big[-\sum_{B \subset \Lambda'} \sigma_\varnothing(B)\, I(B)\big]$ cancels out).
Thus letting

$$\tilde{Z} = \sum_{A \subset \Lambda} \exp\Big[\sum_{B \subset \Lambda'} (\sigma_A(B) - \sigma_\varnothing(B))\, \tilde{I}(B)\Big],$$

we have by Proposition 8.10 that if $E \subset \Lambda$ then

$$Z^{-1} \sum_{A \subset \Lambda} \sigma_E(A) \exp\Big[\sum_{B \subset \Lambda'} (\sigma_A(B) - \sigma_\varnothing(B))\, I(B)\Big]$$
$$\leqslant \tilde{Z}^{-1} \sum_{A \subset \Lambda} \sigma_E(A) \exp\Big[\sum_{B \subset \Lambda'} (\sigma_A(B) - \sigma_\varnothing(B))\, \tilde{I}(B)\Big].$$

But by Proposition 7.6 we have

$$V(A) = \lim_{\Lambda' \to S} \sum_{B \subset \Lambda'} (\sigma_A(B) - \sigma_\varnothing(B))\, I(B),$$

and therefore $\sum_{A \subset \Lambda} \sigma_E(A)\, \pi_\Lambda^\varnothing(A) \leqslant \sum_{A \subset \Lambda} \sigma_E(A)\, \tilde{\pi}_\Lambda^\varnothing(A)$.

Now it is easy to see that

$$\sum_{A \subset \Lambda} \sigma_{\{x\}}(A)\, \pi_\Lambda^\varnothing(A) = 1 - 2 \sum_{x \in A \subset \Lambda} \pi_\Lambda^\varnothing(A) = 1 - 2\rho_\Lambda^\varnothing(\{x\}).$$

Therefore $\qquad 1 - 2\rho_\Lambda^\varnothing(\{x\}) \leqslant 1 - 2\tilde{\rho}_\Lambda^\varnothing(\{x\}),$

i.e. $\qquad \rho_\Lambda^\varnothing(\{x\}) \geqslant \tilde{\rho}_\Lambda^\varnothing(\{x\})$. □

We will now apply this last result to Ising potentials. Recall that by Proposition 7.8 we have that V is an Ising potential if and only if $I_V(E) = 0$ for all $E \in \mathscr{C}(S)$ with $|E|$ odd.

THEOREM 8.4 Let $V, \tilde{V}$ be supermodular Ising potentials with $I_{\tilde{V}}(A) \geqslant I_V(A) \geqslant 0$ for all $A \in \mathscr{C}(S)$; suppose that phase transition occurs for V. Then phase transition occurs for $\tilde{V}$.

Proof By Proposition 8.12 we have that

$$\rho^-(\{x\}) \geqslant \tilde{\rho}^-(\{x\}) \quad \text{for all} \quad x \in S.$$

But by Proposition 8.9 we have that there exists $x \in S$ such that $\rho^-(\{x\}) < \frac{1}{2}$; thus $\tilde{\rho}^-(\{x\}) < \frac{1}{2}$ and so phase transition occurs for $\tilde{V}$. □

NOTES The technique for reducing the problem of phase transition for supermodular potentials to the study of $\rho^+(\{x\})$ and $\rho^-(\{x\})$ is due to Lebowitz and Martin-Löf (1972). The development of various forms of the Griffiths inequalities can be traced by looking at Griffiths (1967), Kelly and Sherman (1968), Sherman (1969), and Ginibre (1970); the proof used here is due to Ginibre.

9. *Attractive pair potentials*

We now specialize still further and look at attractive pair potentials. Thus we consider $V \in H(S)$ with $J_V(B) = 0$ for all $B \in \mathscr{C}(S)$ with $|B| \geqslant 3$ and $J_V(A) \geqslant 0$ for all $A \in \mathscr{C}(S)$ with $|A| = 2$. As in Chapter 7 let $U: S \times S \to \mathbb{R}$ be the bilinear form associated with V, i.e. U is defined by

$$U(x,y) = \begin{cases} \tfrac{1}{2} J_V(\{x,y\}) & \text{if } x \neq y, \\ J_V(\{x\}) & \text{if } x = y. \end{cases}$$

Then for any $A \in \mathscr{C}(S)$ we have $V(A) = U(A,A)$, where for any $A, B \in \mathscr{C}(S)$ we define

$$U(A,B) = \sum_{x \in A} \sum_{y \in B} U(x,y).$$

It is easily checked that for all $x \in S$ we have

$$\sum_{y \in S} J_V(\{x,y\}) < \infty,$$

and thus if $A \in \mathscr{C}(S)$, $X \in \mathscr{P}(S)$ then $U(A,X)$ is well-defined by the above formula. Recall that $\bar{V}$ is also a pair potential and if $\bar{U}$ is the bilinear form associated with $\bar{V}$ then

$$\bar{U}(x,y) = \begin{cases} U(x,y) & \text{if } x \neq y, \\ U(x,x) - 2U(x,S) & \text{if } x = y. \end{cases}$$

In particular V is an Ising potential if and only if $U(x,S) = 0$ for all $x \in S$. Note that in the present set up I_V is certainly defined and is given by

$$I_V(A) = 0 \quad \text{if} \quad A \in \mathscr{C}(S) \quad \text{with } |A| \geqslant 3,$$

$$I_V(\{x,y\}) = \begin{cases} U(x,y) & \text{if } x \neq y, \\ -U(x,S) & \text{if } x = y, \end{cases}$$

$$I_V(\varnothing) = 0.$$

The main technique that we will use in this chapter is the application of the following form of the *Lee–Yang circle theorem*:

THEOREM 9.1 Let $\Lambda = \{1, \ldots, n\}$ and let $(B_{ij})_{1 \leqslant i, j \leqslant n}$ be a symmetric $n \times n$ matrix with $0 < B_{ij} \leqslant 1$ for all i, j. Define a polynomial $p(z_1, \ldots, z_n)$ in the n complex variables $z_1, \ldots, z_n$ by

$$p(z_1, \ldots, z_n) = \sum_{A \subset \Lambda} z^A C(A),$$

where $$z^A = \prod_{i \in A} z_i \quad \text{if} \quad A \neq \varnothing, \quad z^\varnothing = 1,$$

$$C(A) = \prod_{i \in A} \prod_{j \in \Lambda - A} B_{ij} \quad \text{if} \quad A \subset \Lambda \quad \text{with} \quad \varnothing \neq A \neq \Lambda,$$

and $$C(\varnothing) = C(\Lambda) = 1.$$

Let $\xi_1, \ldots, \xi_n \in \mathbb{C}$ with $|\xi_i| \geqslant 1$ for $i = 1, \ldots, n-1$ and

$$p(\xi_1, \ldots, \xi_n) = 0.$$

Then $|\xi_n| \leqslant 1$.

*Proof** By continuity we can assume that $B_{ij} < 1$ for all i, j. We proceed by induction on n; thus to show the dependence on n let us write Λ_n for Λ, p_n for p, and C_n for C. If $n = 1$ then the result is obviously true since $p_1(z_1) = 1 + z_1$. Let $n \geqslant 2$ and suppose that the result is true for all $m < n$. We will break the proof up into a number of Lemmas. (For all the lemmas we make the implicit assumption that the result is true for all $m < n$.)

LEMMA 9.1

Let $s_1, \ldots, s_{n-1} \in \mathbb{C}$ with $|s_i| = 1$ for $i = 1, \ldots, n-1$.

Let $\xi \in \mathbb{C}$ and suppose that $p_n(s_1, \ldots, s_{n-1}, \xi) = 0$. Then $|\xi| = 1$.

Proof Since $|s_i| = 1$ we have $s_i^{-1} = \bar{s}_i$ (where $\bar{z}$ denotes the complex conjugate of z); and since $B_{ij} = B_{ji}$ we have for all non-zero $z_1, \ldots, z_n \in \mathbb{C}$ that

$$p_n(z_1^{-1}, \ldots, z_n^{-1}) = z_1^{-1}, \ldots, z_n^{-1} p_n(z_1, \ldots, z_n).$$

* For another proof of the theorem see the Appendix.

Thus if $\xi \neq 0$ we have

$$\begin{aligned} p_n(s_1, \ldots, s_{n-1}, (\bar{\xi})^{-1}) &= p_n((\bar{s}_1)^{-1}, \ldots, (\bar{s}_{n-1})^{-1}, (\bar{\xi})^{-1}) \\ &= (\bar{s}_1)^{-1} \ldots (\bar{s}_{n-1})^{-1} (\bar{\xi})^{-1} p_n(\bar{s}_1, \ldots, \bar{s}_{n-1}, \bar{\xi}) \\ &= (\bar{s}_1)^{-1} \ldots (\bar{s}_{n-1})^{-1} (\bar{\xi})^{-1} \overline{p_n(s_1, \ldots, s_{n-1}, \xi)} = 0. \end{aligned}$$

(We have $p_n(\bar{z}_1, \ldots, \bar{z}_n) = \overline{p_n(z_1, \ldots, z_n)}$ as p_n has real coefficients.) Now we can write

$$p_n(s_1, \ldots, s_{n-1}, z) = a + bz$$

for some $a, b \in \mathbb{C}$. If we could show that $a \neq 0$ then $\xi \neq 0$, and if also $b \neq 0$ then we have shown that

$$\xi = -\frac{a}{b} = (\bar{\xi})^{-1}$$

and hence $|\xi| = 1$. Therefore the proof of the lemma will be complete if we show that $a \neq 0$ and $b \neq 0$. We have

$$\begin{aligned} a &= \sum_{A \subset \Lambda_{n-1}} s^A \Big(\prod_{i \in A} \prod_{j \in \Lambda_n - A} B_{ij} \Big) \\ &= \sum_{A \subset \Lambda_{n-1}} s^A \Big(\prod_{i \in A} B_{in} \Big) \Big(\prod_{i \in A} \prod_{j \in \Lambda_{n-1} - A} B_{ij} \Big) \\ &= p_{n-1}(B_{1n} s_1, \ldots, B_{n-1,n} s_{n-1}) \\ &= \prod_{i=1}^{n-1} B_{in} s_i p_{n-1}(B_{1n}^{-1} s_1^{-1}, \ldots, B_{n-1,n}^{-1} s_{n-1}^{-1}) \neq 0 \end{aligned}$$

by the induction hypothesis since for $i = 1, \ldots, n-1$ we have

$$|B_{1n}^{-1} s_1^{-1}| = |B_{1n}^{-1}| > 1.$$

Similarly we have

$$\begin{aligned} b &= \sum_{A \subset \Lambda_{n-1}} s^A \Big(\prod_{i \in A \cup \{n\}} \prod_{j \in \Lambda_{n-1} - A} B_{ij} \Big) \\ &= \sum_{A \subset \Lambda_{n-1}} s^A \Big(\prod_{i \in A} \prod_{j \in \Lambda_{n-1} - A} B_{ij} \Big) \left(\prod_{i=1}^{n-1} B_{ni} \right) \Big(\prod_{i \in A} B_{ni}^{-1} \Big) \\ &= \left(\prod_{i=1}^{n-1} B_{ni} \right) p_{n-1}(B_{n1}^{-1} s_1, \ldots, B_{n,n-1}^{-1} s_{n-1}) \neq 0 \end{aligned}$$

since $\quad |B_{ni}^{-1} s_i| > 1 \quad \text{for} \quad i = 1, \ldots, n-1.$ □

LEMMA 9.2

Let $\xi_1, \ldots, \xi_{n-2} \in \mathbb{C}$ with $|\xi_i| \geqslant 1$ for $i = 1, \ldots, n-2$. We can write

$$p_n(\xi_1, \ldots, \xi_{n-2}, w, z) = \alpha + \beta z + \gamma w + \delta zw.$$

Then $\delta \neq 0$.

Proof We have

$$\delta = \sum_{A \subset \Lambda_{n-2}} \xi^A \Big(\prod_{i \in A \cup \{n-1, n\}} \prod_{j \in \Lambda_{n-2} - A} B_{ij} \Big)$$

$$= \left(\prod_{j=1}^{n-2} B_{n-1,j} B_{n,j} \right) p_{n-2}(\eta_1, \ldots, \eta_{n-2})$$

where $\eta_i = (B_{n-1,j} B_{n,j})^{-1} \xi_i$ for $i = 1, \ldots, n-2$. But

$$|\eta_i| > |\xi_i| \geqslant 1 \quad \text{for} \quad i = 1, \ldots, n-2$$

and thus by the induction hypothesis we have $\delta \neq 0$.□

LEMMA 9.3

Let $\xi_1, \ldots, \xi_{n-2} \in \mathbb{C}$ with $|\xi_i| \geqslant 1$ for $i = 1, \ldots, n-2$, and again write

$$p_n(\xi_1, \ldots, \xi_{n-2}, w, z) = \alpha + \beta z + \gamma w + \delta zw.$$

Then

$$\left| \frac{\beta}{\delta} \right| < 1.$$

Proof Let $\xi_{n-1} = -\beta/\delta$; then we have

$$0 = \beta + \xi_{n-1}\delta = \left(\prod_{i=1}^{n-1} B_{ni} \right) p_{n-1}(B_{n1}^{-1}\xi_1, \ldots, B_{n,n-1}^{-1}\xi_{n-1}).$$

But $|B_{ni}^{-1}\xi_i| > 1$ for $i = 1, \ldots, n-2$ and thus by the induction hypothesis we have $|B_{n,n-1}^{-1}\xi_{n-1}| \leqslant 1$. Therefore $|\xi_{n-1}| < 1$, i.e. $|\beta/\delta| < 1$.□

LEMMA 9.4

Let $\xi_1, \ldots, \xi_{n-2} \in \mathbb{C}$ with $|\xi_i| \geqslant 1$ for $i = 1, \ldots, n-2$,

and let $w', z' \in \mathbb{C}$ with $|w'| > 1$, $|z'| > 1$ and suppose

$$p_n(\xi_1, \ldots, \xi_{n-2}, w', z') = 0.$$

Then there exist $w'', z'' \in \mathbb{C}$ with $|w''| = 1$, $|z''| > 1$ and

$$p_n(\xi_1, \ldots, \xi_{n-2}, w'', z'') = 0.$$

Proof As before write

$$p_n(\xi_1, \ldots, \xi_{n-2}, w, z) = \alpha + \beta z + \gamma w + \delta z w.$$

Then the equation

$$p_n(\xi_1, \ldots, \xi_{n-2}, w, z) = 0$$

gives us a fractional linear transformation

$$w = T(z) = -\frac{\alpha + \beta z}{\gamma + \delta z}.$$

Now T maps $z = \infty$ to $w = -\beta/\delta$ and from Lemma 9.3 we have $|-\beta/\delta| < 1$. Also of course T maps z' to w'. Thus by continuity there exists $z'' \in \mathbb{C}$ with $|z''| > 1$ and $|T(z'')| = 1$. (In fact we can take z'' to be of the form tz' for some $t > 1$.) Let $w'' = T(z'')$, thus $p_n(\xi_1, \ldots, \xi_{n-2}, w'', z'') = 0.$ □

We are now in a position to complete the proof of the theorem. Suppose the theorem is not true for n; then there exist

$$\xi_1, \ldots, \xi_{n-2}, w', z' \in \mathbb{C}$$

with $|\xi_i| \geqslant 1$ for $i = 1, \ldots, n-2$, $|w'| > 1$, $|z'| > 1$ and

$$p_n(\xi_1, \ldots, \xi_{n-2}, w', z') = 0.$$

Therefore the hypotheses of Lemma 9.4 are satisfied, thus there exist $s_1, w_1 \in \mathbb{C}$ with $|s_1| = 1$, $|w_1| > 1$ and

$$p_n(\xi_1, \ldots, \xi_{n-2}, s_1, w_1) = 0.$$

Now p_n is symmetric in its entries, thus

$$p_n(s_1, \xi_1, \ldots, \xi_{n-2}, w_1) = 0.$$

If $|\xi_{n-2}| = 1$ then we let $s_2 = \xi_{n-2}$, if $|\xi_{n-2}| > 1$ then we can apply Lemma 9.4 again to get $s_2, w_2, \in \mathbb{C}$ with $|s_2| = 1, |w_2| > 1$ and $p_n(s_1, \xi_1, \ldots, s_2, w_2) = 0$. Therefore in any case we have

$$p_n(s_1, s_2, \xi_1, \ldots, \xi_{n-3}, w_2) = 0$$

for some $s_2, w_2 \in \mathbb{C}$ with $|s_2| = 1$, $|w_2| > 1$. It is clear that we can repeat this procedure until we end up with $s_1, \ldots, s_{n-1}, \xi \in \mathbb{C}$, $|s_i| = 1$ for $i = 1, \ldots, n-1$, $|\xi| > 1$, and $p_n(s_1, \ldots, s_{n-1}, \xi) = 0$. But this contradicts Lemma 9.1; and thus the proof of the theorem is complete. □

We will investigate the properties of the pair potential V by looking at a family of pair potentials parametrized by a variable λ and seeing how the correlation functions behave as a function of λ. Let $W: S \to \mathbb{R}$ with $W(x) \geqslant 0$ for all $x \in S$, and for $\lambda \in \mathbb{R}$ let V_λ be the pair potential with associated bilinear form U_λ, where

$$U_\lambda(x, y) = \begin{cases} U(x, y) & \text{if} \quad x \neq y, \\ U(x, x) + \lambda W(x) & \text{if} \quad x = y. \end{cases}$$

Since we are assuming that V is attractive, i.e. $U(x, y) \geqslant 0$ if $x \neq y$, then it is clear that V_λ is also attractive. If $\Lambda \in \mathscr{C}(S)$ and $X \in \mathscr{P}(S - \Lambda)$ then we let $\pi^X_{\Lambda,\lambda}$ denote the Gibbs state on Λ with potential V_λ and boundary condition X. Let $\rho^X_{\Lambda,\lambda}: \mathscr{C}(\Lambda) \to \mathbb{R}$ be the correlation function of $\pi^X_{\Lambda,\lambda}$. Note that we can write

$$\rho^X_{\Lambda,\lambda}(A) = (Z^X_{\Lambda,\lambda})^{-1} \sum_{A \subset B \subset \Lambda} \exp\left[\lambda W(B) + U(B, B) + 2U(B, X)\right],$$

where $\quad Z^X_{\Lambda,\lambda} = \sum_{B \subset \Lambda} \exp\left[\lambda W(B) + U(B, B) + 2U(B, X)\right];$

and we can thus think of $\rho^X_{\Lambda,\lambda}(A)$ as a meromorphic function of the complex variable λ.

Proposition 9.1 Let $\Lambda \in \mathscr{C}(S)$, $X \in \mathscr{P}(S - \Lambda)$ and $x \in \Lambda$. Suppose that for some $\epsilon > 0$ we have

$$U(y, \Lambda) + 2U(y, X) - W(y) \geqslant \epsilon \quad \text{for all} \quad y \in \Lambda.$$

Let $f(\lambda) = \rho^X_{\Lambda,\lambda}(\{x\})$; then f is holomorphic in $\{z: \operatorname{Re} z > -1\}$ and in fact $\quad |f(z)| \leqslant [1 - \exp(-\epsilon)]^{-1} \quad \text{if} \quad \operatorname{Re} z > -1.$

Proof Let $z_0 \in \mathbb{C}$ and suppose that $f(\lambda) = z_0$ for some λ with $\operatorname{Re} \lambda > -1$ and λ not a zero of $Z^X_{\Lambda,\lambda}$. Then we have

$$z_0^{-1} \sum_{x \in A \subset \Lambda} \exp\left[\lambda W(A) + U(A, A) + 2U(A, X)\right]$$
$$= \sum_{B \subset \Lambda} \exp\left[\lambda W(B) + U(B, B) + 2U(B, X)\right],$$

and thus

$$\sum_{A \subset \Lambda} \gamma(A) \exp[\lambda W(A) + U(A,A) + 2U(A,X)] = 0,$$

where
$$\gamma(A) = \begin{cases} 1 & \text{if } x \notin A, \\ \left(1 - \dfrac{1}{z_0}\right) & \text{if } x \in A. \end{cases}$$

Now we have

$$\begin{aligned} &\lambda W(A) + U(A,A) + 2U(A,X) \\ &\quad = \lambda W(A) + U(A,\Lambda) + 2U(A,X) - U(A, \Lambda - A) \\ &\quad = \sum_{y \in A} [\lambda W(y) + U(y,\Lambda) + 2U(y,X)] - \sum_{i \in A} \sum_{j \in \Lambda - A} U(i,j). \end{aligned}$$

Define $h(\cdot, \lambda) : \Lambda \to \mathbb{C}$ by

$$h(y, \lambda) = \lambda W(y) + U(y, \Lambda) + 2U(y, X)$$

and $\bar{\gamma} : \Lambda \to \mathbb{C}$ by

$$\bar{\gamma}(y) = \begin{cases} 1 & \text{if } y \neq x, \\ \left(1 - \dfrac{1}{z_0}\right) & \text{if } y = x. \end{cases}$$

Then we have

$$\begin{aligned} 0 &= \sum_{A \subset \Lambda} \gamma(A) \exp[\lambda W(A) + U(A,A) + 2U(A,X)] \\ &= \sum_{A \subset \Lambda} \Big(\prod_{y \in A} \bar{\gamma}(y) \exp h(y, \lambda) \Big) \Big(\prod_{i \in A} \prod_{j \in \Lambda - A} \exp(-U(i,j)) \Big). \end{aligned}$$

Note that (for $i \neq j$) we have $0 < \exp(-U(i,j)) \leqslant 1$, and thus by Theorem 9.1 we cannot have $f(\lambda) = z_0$ if $|\bar{\gamma}(y) \exp h(y, \lambda)| > 1$ for all $y \in \Lambda$. But if $|z_0| > [1 - \exp(-\epsilon)]^{-1}$ then it is easy to check that $|\bar{\gamma}(y) \exp h(y, \lambda)| > 1$ for all $y \in \Lambda$. Therefore we have

$$|f(z)| \leqslant [1 - \exp(-\epsilon)]^{-1}$$

if $\operatorname{Re} z > -1$ and if z is not a zero of $Z^X_{\Lambda,\lambda}$. Now it must clearly still hold if z is a zero of $Z^X_{\Lambda,\lambda}$ (with $\operatorname{Re} z > -1$), hence f is holomorphic in $\{z : \operatorname{Re} z > -1\}$ and

$$|f(z)| \leqslant [1 - \exp(-\epsilon)]^{-1} \quad \text{if} \quad \operatorname{Re} z > -1. \square$$

For $\lambda \in \mathbb{R}$ let $\rho^+_{\Lambda,\lambda} = \rho^{S-\Lambda}_{\Lambda,\lambda}$, $\rho^-_{\Lambda,\lambda} = \rho^{\varnothing}_{\Lambda,\lambda}$, and let

$$\rho^+_\lambda = \lim_{\Lambda \uparrow S} \rho^+_{\Lambda,\lambda}, \quad \rho^-_\lambda = \lim_{\Lambda \uparrow S} \rho^-_{\Lambda,\lambda}.$$

We now have

Proposition 9.2 Suppose there exist $\epsilon > 0$ and $\Lambda_n \uparrow S$ such that

$$U(y, \Lambda_n) - W(y) \geq \epsilon \quad \text{for all} \quad y \in \Lambda_n \quad \text{and all} \quad n.$$

Let $x \in S$; then there exist f^+, f^-, holomorphic in $\{z: \operatorname{Re} z > -1\}$ such that for all $\lambda \in \mathbb{R}$ with $\lambda > -1$ we have $f^+(\lambda) = \rho^+_\lambda(\{x\})$ and $f^-(\lambda) = \rho^-_\lambda(\{x\})$.

Proof Let $f^+_n(\lambda) = \rho^{S-\Lambda_n}_{\Lambda_n,\lambda}(\{x\})$, then by Proposition 9.1 we have that f^+_n is holomorphic in $\{z: \operatorname{Re} z > -1\}$ and

$$|f^+_n(z)| \leq [1 - \exp(-\epsilon)]^{-1} \quad \text{if} \quad \operatorname{Re} z > -1.$$

Therefore $\{f^+_n\}_{n=1}^\infty$ forms a normal family and thus there exists f^+ holomorphic in $\{z: \operatorname{Re} z > -1\}$ and a subsequence $\{n_j\}$ such that $f^+_{n_j} \to f^+$ as $j \to \infty$ (with uniform convergence on compact subsets of $\{z: \operatorname{Re} z > -1\}$). But for all $\lambda \in \mathbb{R}$ we have

$$f^+_n(\lambda) \to \rho^+_\lambda(\{x\}) \quad \text{as} \quad n \to \infty,$$

and thus $f^+(\lambda) = \rho^+_\lambda(\{x\})$ for $\lambda > -1$. Clearly the same proof works for $\rho^-_\lambda(\{x\})$. □

For $\Lambda \in \mathscr{C}(S)$ let $m(\Lambda, U) = \max_{y \in \Lambda} U(y, S - \Lambda)$,

and let $$m^*(U) = \liminf_{\Lambda \uparrow S} m(\Lambda, U).$$

Theorem 9.2 Suppose there exists $\epsilon > 0$ such that either:

(i) $U(x, S) \geq m^*(U) + \epsilon(1 + |U(x, x)|)$ for all $x \in S$, or

(ii) $-U(x, S) \geq m^*(U) + \epsilon(1 + |U(x, x)|)$ for all $x \in S$.

Then phase transition cannot occur for V.

Proof Suppose that (i) holds. Define $W: S \to \mathbb{R}$ by

$$W(y) = \epsilon(\tfrac{1}{3} + |U(y, y)|).$$

Now by Theorem 8.2 (applied to $\bar{V}_\lambda$) we have that phase transition cannot occur for V_λ if

$$\inf_{x \in S} [\exp V_\lambda(\{x\}) (1 + \exp \bar{V}_\lambda(\{x\}))] > 1,$$

and thus it is easily computed that phase transition cannot occur for V_λ if $\lambda \geqslant 1/\epsilon$. Since (i) holds there exist $\Lambda_n \in \mathscr{C}(S)$ with $\Lambda_n \uparrow S$ and such that

$$U(x,S) \geqslant m(\Lambda_n, U) - \tfrac{1}{3}\epsilon + \epsilon(1 + |U(x,x)|) \quad \text{for all} \quad x \in S,$$

and thus

$$U(y,S) \geqslant U(y, S - \Lambda_n) + \tfrac{1}{3}\epsilon + W(y) \quad \text{for all} \quad y \in \Lambda_n,$$

i.e.

$$U(y, \Lambda_n) - W(y) \geqslant \tfrac{1}{3}\epsilon \quad \text{for all} \quad y \in \Lambda_n.$$

Therefore by Proposition 9.2 we have that if $x \in S$ then there exist f^+, f^- holomorphic in $\{z : \operatorname{Re} z > -1\}$ such that for all $\lambda \in \mathbb{R}$ with $\lambda > -1$ we have $f^+(\lambda) = \rho_\lambda^+(\{x\})$ and $f^-(\lambda) = \rho_\lambda^-(\{x\})$. But if $\lambda \in \mathbb{R}$ with $\lambda \geqslant 1/\epsilon$ then as phase transition does not occur for V_λ we must have $\rho_\lambda^+(\{x\}) = \rho_\lambda^-(\{x\})$. Thus $f^+(\lambda) = f^-(\lambda)$ for all $\lambda \in \mathbb{R}$ with $\lambda \geqslant 1/\epsilon$ and hence by the uniqueness theorem for holomorphic functions we have $f^+(z) = f^-(z)$ for all z with $\operatorname{Re} z > -1$. In particular we have $f^+(0) = f^-(0)$, i.e.

$$\rho_0^+(\{x\}) = \rho_0^-(\{x\}).$$

This is true for all $x \in S$, thus by Theorem 8.1 phase transition cannot occur for V. The other half of the theorem is proved in the same way using $\bar{V}$ instead of V. □

Recall that V is an Ising potential if and only if $U(x,S) = 0$ for all $x \in S$. Thus Theorem 9.2 roughly says that phase transition does not occur for potentials which are not close to being Ising potentials. If we put more conditions on the potential V then we can improve on Theorem 9.2. Let V_λ be as before (with W any function with $W(x) \geqslant 0$ for all $x \in S$); for $\Lambda \in \mathscr{C}(S)$ and $\lambda \in \mathbb{R}$ define $\pi^0_{\Lambda,\lambda}$ to be the element of $\mathscr{S}(\Lambda)$ with density given by

$$\pi^0_{\Lambda,\lambda}(A) = (Z^0_{\Lambda,\lambda})^{-1} \exp [\lambda W(A) + U(A,A) + U(A, S - \Lambda)],$$

where

$$Z^0_{\Lambda,\lambda} = \sum_{B \subset \Lambda} \exp[\lambda W(B) + U(B,B) + U(B, S-\Lambda)].$$

Let $\rho^0_{\Lambda,\lambda}$ denote the correlation function of $\pi^0_{\Lambda,\lambda}$; we will write ρ^0_Λ instead of $\rho^0_{\Lambda,0}$. It is easily seen using Holley's inequality that

$$\rho^-_\Lambda \leqslant \rho^0_\Lambda \leqslant \rho^+_\Lambda.$$

PROPOSITION 9.3 Suppose given any $x \in S$ there exist α with $0 \leqslant \alpha < 1$ and $\Lambda_n \uparrow S$ such that for all n we have

$$\begin{aligned}(1-\alpha)\rho^+_{\Lambda_n}(\{x\}) + \alpha\rho^-_{\Lambda_n}(\{x\}) &\leqslant \rho^0_{\Lambda_n}(\{x\}) \\ &\leqslant \alpha\rho^+_{\Lambda_n}(\{x\}) + (1-\alpha)\rho^-_{\Lambda_n}(\{x\}).\end{aligned}$$

If $\epsilon > 0$ then phase transition cannot occur for V if either

(i) $U(x,S) \geqslant \epsilon(1+|U(x,x)|)$ for all $x \in S$, or

(ii) $-U(x,S) \geqslant \epsilon(1+|U(x,x)|)$ for all $x \in S$.

Proof This is much the same as the proof of Theorem 9.2. Considering $\rho^0_{\Lambda,\lambda}(\{x\})$ as a meromorphic function of λ we can show that if (i) holds then just as in Propositions 9.1, 9.2 and Theorem 9.2 we have $\rho^0(\{x\}) = \rho^+(\{x\})$. But we clearly have

$$\begin{aligned}(1-\alpha)\rho^+(\{x\}) + \alpha\rho^-(\{x\}) &\leqslant \rho^0(\{x\}) \\ &\leqslant \alpha\rho^+(\{x\}) + (1-\alpha)\rho^-(\{x\})\end{aligned}$$

and thus we get $\rho^+(\{x\}) = \rho^-(\{x\})$. Therefore phase transition cannot occur for V. □

Unfortunately it is not very easy to check when the hypothesis of Proposition 9.3 is satisfied. There is however one case when it can be done. If $\Lambda \in \mathscr{C}(S)$ then define

$$Z^+_\Lambda = \sum_{A \subset \Lambda} \exp[U(A,A) + 2U(A, S-\Lambda)],$$

$$Z^-_\Lambda = \sum_{A \subset \Lambda} \exp U(A,A).$$

PROPOSITION 9.4 Let $x \in \Lambda \in \mathscr{C}(S)$ and $\alpha \in \mathbb{R}$ with

$$0 < \alpha < 1.$$

Then

(i) If
$$\frac{\alpha}{1-\alpha} \geqslant Z_\Lambda^+/Z_\Lambda^-$$

then
$$\rho_\Lambda^0(\{x\}) \leqslant \alpha\rho_\Lambda^+(\{x\}) + (1-\alpha)\,\rho_\Lambda^-(\{x\}).$$

(ii) If
$$\frac{1-\alpha}{\alpha} \geqslant [Z_\Lambda^- \exp 2U(\Lambda, S-\Lambda)]/Z_\Lambda^+$$

then
$$\alpha\rho_\Lambda^+(\{x\}) + (1-\alpha)\,\rho_\Lambda^-(\{x\}) \leqslant \rho_\Lambda^0(\{x\}).$$

Proof Let $\mu_1, \mu_2 \in \mathscr{S}(\Lambda)$ be given by

$$\mu_1(A) = \frac{\alpha}{Z_\Lambda^+}\exp[U(A,A) + 2U(A, S-\Lambda)] + \frac{(1-\alpha)}{Z_\Lambda^-}\exp U(A,A),$$

$$\mu_2(A) = \frac{1}{Z_\Lambda^0}\exp[U(A,A) + U(A, S-\Lambda)] \quad \text{for} \quad A \subset \Lambda.$$

Then we have

$$\sum_{x \in A \subset \Lambda} \mu_1(A) = \alpha\rho_\Lambda^+(\{x\}) + (1-\alpha)\,\rho_\Lambda^-(\{x\}),$$

$$\sum_{x \in A \subset \Lambda} \mu_2(A) = \rho_\Lambda^0(\{x\}).$$

If
$$\frac{\alpha}{1-\alpha} \geqslant Z_\Lambda^+/Z_\Lambda^-$$

then we leave it to the reader to check that for all $A, B \subset \Lambda$ we have
$$\mu_1(A \cup B)\,\mu_2(A \cap B) \geqslant \mu_1(A)\,\mu_2(B).$$

Therefore by Holley's inequality we have

$$\rho_\Lambda^0(\{x\}) \leqslant \alpha\rho_\Lambda^+(\{x\}) + (1-\alpha)\,\rho_\Lambda^-(\{x\}).$$

The proof of (ii) follows in the same way. □

By Proposition 9.4 we have that V satisfies the hypothesis of Proposition 9.3 if
$$\lim_{\Lambda \uparrow S} U(\Lambda, S-\Lambda) < \infty.$$

We will now use a different approach to improve on Theorem 9.2 for certain potentials. Again let $W: S \to \mathbb{R}$ with $W(x) \geqslant 0$ for all $x \in S$, and suppose that $W(y) > 0$ for some $y \in S$. We continue to use the notation that for $\Lambda \in \mathscr{C}(S)$,

$$W(\Lambda) = \sum_{x \in \Lambda} W(x);$$

if $\Lambda \in \mathscr{C}(S)$ and $X \in \mathscr{P}(S-\Lambda)$ then let $Z^X_{\Lambda,\lambda}$ and $\rho^X_{\Lambda,\lambda}$ be as before. We define $P^X_\Lambda : \mathbb{R} \to \mathbb{R}$ by

$$P^X_\Lambda(\lambda) = \frac{1}{W(\Lambda)} \log Z^X_{\Lambda,\lambda}$$

(since we will only be interested in what happens when $\Lambda \uparrow S$ we will not worry about $W(\Lambda)$ being zero). The following properties of P^X_Λ are not difficult to verify:

PROPOSITION 9.5

(i) $$\frac{\mathrm{d}}{\mathrm{d}\lambda}(P^X_\Lambda(\lambda)) = \frac{1}{W(\Lambda)} \sum_{x \in \Lambda} W(x)\, \rho^X_{\Lambda,\lambda}(\{x\}).$$

(ii) P^X_Λ is a convex function of λ.

(iii) If $X, Y \in \mathscr{P}(S-\Lambda)$ then

$$|P^X_\Lambda(\lambda) - P^Y_\Lambda(\lambda)| \leqslant \frac{2U(\Lambda, S-\Lambda)}{W(\Lambda)}.$$

Proof (i) We have

$$\begin{aligned}
\frac{\mathrm{d}}{\mathrm{d}\lambda}(P^X_\Lambda(\lambda)) &= \frac{1}{W(\Lambda)} \sum_{A \subset \Lambda} W(A)\, \pi^X_{\Lambda,\lambda}(A) \\
&= \frac{1}{W(\Lambda)} \sum_{A \subset \Lambda} \sum_{x \in A} W(x)\, \pi^X_{\Lambda,\lambda}(A) \\
&= \frac{1}{W(\Lambda)} \sum_{x \in \Lambda} \sum_{x \in A \subset \Lambda} W(x)\, \pi^X_{\Lambda,\lambda}(A) \\
&= \frac{1}{W(\Lambda)} \sum_{x \in \Lambda} W(x)\, \rho^X_{\Lambda,\lambda}(\{x\}).
\end{aligned}$$

(ii) Let $\lambda_1, \lambda_2 \in \mathbb{R}$ with $\lambda_1 < \lambda_2$ and let $0 < t < 1$. Then

$$\begin{aligned}
&P^X_\Lambda(t\lambda_1 + (1-t)\lambda_2) \\
&= \frac{1}{W(\Lambda)} \log \{ \sum_{B \subset \Lambda} \exp [(t\lambda_1 + (1-t)\lambda_2)\, W(B) + U(B,B) \\
&\qquad\qquad + 2U(B,X)]\}
\end{aligned}$$

$$= \frac{1}{W(\Lambda)} \log\{\sum_{B\subset\Lambda} [\exp[\lambda_1 W(B) + U(B,B) + 2U(B,X)]]^t$$
$$\times [\exp[\lambda_2 W(B) + U(B,B) + 2U(B,X)]]^{(1-t)}\}$$
$$\leqslant \frac{1}{W(\Lambda)} \log\{[\sum_{B\subset\Lambda} \exp[\lambda_1 W(B) + U(B,B) + 2U(B,X)]]^t$$
$$\times [\sum_{B\subset\Lambda} \exp[\lambda_2 W(B) + U(B,B) + 2U(B,X)]]^{(1-t)}\}$$

(by Holder's inequality)

$$= tP_\Lambda^X(\lambda_1) + (1-t)\, P_\Lambda^X(t_2).$$

Therefore P_Λ^X is a convex function of λ.

(iii) This is a simple calculation. □

In order to prove the next theorem we need the following result from real analysis:

LEMMA 9.5 Let $I \subset \mathbb{R}$ be an open interval and for $n = 1, 2, \ldots$ let $f_n : I \to \mathbb{R}$ be convex and differentiable at some point $x_0 \in I$. Suppose that $f_n(x) \to f(x)$ as $n \to \infty$ for all $x \in I$ and that $f: I \to \mathbb{R}$ is also differentiable at x_0. Then

$$f'(x_0) = \lim_{n\to\infty} f_n'(x_0).$$

Proof Define $g_n : I - \{x_0\} \to \mathbb{R}$ by

$$g_n(x) = \frac{f_n(x) - f_n(x_0)}{x - x_0};$$

and similarly $g: I - \{x_0\} \to \mathbb{R}$ by

$$g(x) = \frac{f(x) - f(x_0)}{x - x_0}.$$

Thus $g_n(x) \to g(x)$ as $n \to \infty$ if $x \neq x_0$. Now as f_n is conve must have

$$f_n'(x_0) = \inf_{x > x_0} g_n(x) = \sup_{x < x_0} g_n(x);$$

and since f must also be convex we have

$$f'(x_0) = \inf_{x > x_0} g(x) = \sup_{x < x_0} g(x).$$

Therefore $\qquad f_n'(x_0) \leqslant g_n(x) \quad \text{for all} \quad x > x_0,$

and thus

$$\limsup_{n\to\infty} f_n'(x_0) \leqslant \lim_{n\to\infty} g_n(x) = g(x) \quad \text{for all} \quad x > x_0.$$

Hence
$$\limsup_{n\to\infty} f_n'(x_0) \leqslant \inf_{x>x_0} g(x) = f'(x_0).$$

On the other hand since $f_n'(x_0) \geqslant g_n(x)$ for all $x < x_0$ we must have

$$\liminf_{n\to\infty} f_n'(x_0) \geqslant f'(x_0).$$

Thus $\lim_{n\to\infty} f_n'(x_0)$ exists and equals $f'(x_0)$. □

THEOREM 9.3 Suppose that

$$\lim_{\Lambda\uparrow S} \frac{1}{W(\Lambda)} \sum_{x\in\Lambda} W(x)\,\rho^+(\{x\})$$

and
$$\lim_{\Lambda\uparrow S} \frac{1}{W(\Lambda)} \sum_{x\in\Lambda} W(x)\,\rho^-(\{x\})$$

exist; and denote the limits by W^+ and W^- respectively. Suppose also that

$$\liminf_{\Lambda\uparrow S} \frac{U(\Lambda, S-\Lambda)}{W(\Lambda)} = 0,$$

and that for some $\epsilon > 0$ either:

(i) $U(x,S) \geqslant \epsilon(1+W(x))$ for all $x\in S$, or

(ii) $-U(x,S) \geqslant \epsilon(1+W(x))$ for all $x\in S$.

Then $W^+ = W^-$.

Proof For $\Lambda\in\mathscr{C}(S)$ let us write P_Λ^+ (resp. P_Λ^-) for $P_\Lambda^{S-\Lambda}$ (resp. $P_\Lambda^{\varnothing}$). Let us define f_Λ^+, f_Λ^- by

$$f_\Lambda^+(\lambda) = \frac{1}{W(\Lambda)} \sum_{x\in\Lambda} W(x)\,\rho_{\Lambda,\lambda}^{S-\Lambda}(\{x\});$$

$$f_\Lambda^-(\lambda) = \frac{1}{W(\Lambda)} \sum_{x\in\Lambda} W(x)\,\rho_{\Lambda,\lambda}^{\varnothing}(\{x\}).$$

Suppose that (i) holds, then

$$U(y,\Lambda) + U(y,S-\Lambda) \geqslant \epsilon + \epsilon W(y) \quad \text{for all} \quad y\in\Lambda,$$

thus

$$U(y,\Lambda) + 2U(y,S-\Lambda) - \epsilon W(y) \geqslant \epsilon \quad \text{for all} \quad y\in\Lambda.$$

Therefore by Proposition 9.1 we have that if $x \in \Lambda$ then $\rho^{S-\Lambda}_{\Lambda,\lambda}(\{x\})$ is holomorphic in $\{\lambda : \operatorname{Re}\lambda > -\epsilon\}$ and

$$|\rho^{S-\Lambda}_{\Lambda,\lambda}(\{x\})| \leqslant [1-\exp(-\epsilon)]^{-1} \quad \text{if} \quad \operatorname{Re}\lambda > -\epsilon.$$

Thus we can think of f^+_Λ as being holomorphic in $\{z : \operatorname{Re} z > -\epsilon\}$ and we have

$$|f^+_\Lambda(z)| \leqslant [1-\exp(-\epsilon)]^{-1} \quad \text{if} \quad \operatorname{Re} z > -\epsilon.$$

We therefore have that $\{f^+_\Lambda\}_{\Lambda \in \mathscr{C}(S)}$ is a normal family, hence there exists f^+ holomorphic in $\{z : \operatorname{Re} z > -\epsilon\}$ and $\Lambda_n \uparrow S$ such that

(*a*) $$\lim_{n\to\infty} \frac{U(\Lambda_n, S-\Lambda_n)}{W(\Lambda_n)} = 0,$$

and

(*b*) $f^+_{\Lambda_n} \to f^+$ uniformly on compact subsets of $\{z : \operatorname{Re} z > -\epsilon\}$.

Now define $g^+_n, g^-_n, g : (-\epsilon, \epsilon) \to \mathbb{R}$ by

$$g^+_n(\lambda) = \int_0^\lambda f^+_{\Lambda_n}(t)\,\mathrm{d}t,$$

$$g^-_n(\lambda) = \int_0^\lambda f^-_{\Lambda_n}(t)\,\mathrm{d}t,$$

$$g(\lambda) = \int_0^\lambda f^+(t)\,\mathrm{d}t.$$

Then by Proposition 9.5 (i) we have

$$g^+_n(\lambda) = P^+_\Lambda(\lambda) - P^+_\Lambda(0),$$

$$g^-_n(\lambda) = P^-_\Lambda(\lambda) - P^-_\Lambda(0),$$

and thus g^+_n, g^-_n are convex functions. It is clear that $g^+_n(\lambda) \to g(\lambda)$ for all $\lambda \in (-\epsilon, \epsilon)$, but in fact we also have $g^-_n(\lambda) \to g(\lambda)$ for all $\lambda \in (-\epsilon, \epsilon)$, since by Proposition 9.5 (iii) we have

$$|g^+_n(\lambda) - g^-_n(\lambda)| \leqslant |\lambda|\, 4\, \frac{U(\Lambda_n, S-\Lambda_n)}{W(\Lambda_n)}$$

and thus $g^+_n(\lambda) - g^-_n(\lambda) \to 0$. Therefore by Lemma 9.5 we have

$$\lim_{n\to\infty} f^+_{\Lambda_n}(0) = f^+(0) = \lim_{n\to\infty} f^-_{\Lambda_n}(0).$$

But for any $\Lambda \in \mathscr{C}(S)$ and $x \in \Lambda$ we have

$$\rho_\Lambda^-(\{x\}) \leqslant \rho^-(\{x\}) \leqslant \rho^+(\{x\}) \leqslant \rho_\Lambda^+(\{x\})$$

and thus

$$f_\Lambda^-(0) \leqslant \frac{1}{W(\Lambda)} \sum_{x \in \Lambda} W(x) \rho^-(\{x\}) \leqslant \frac{1}{W(\Lambda)} \sum_{x \in \Lambda} W(x) \rho^+(\{x\}) \leqslant f_\Lambda^+(0).$$

Therefore $$\lim_{n \to \infty} f_{\Lambda_n}^-(0) \leqslant W^- \leqslant W^+ \leqslant \lim_{n \to \infty} f_{\Lambda_n}^+(0),$$

i.e. $W^+ = W^-$. Finally if (ii) holds rather than (i) then (i) holds for $\bar{V}$ and the same proof still works since we have

$$\bar{\rho}^+(\{x\}) = 1 - \rho^-(\{x\}) \quad \text{and} \quad \bar{\rho}^-(\{x\}) = 1 - \rho^+(\{x\}). \square$$

If $$\liminf_{\Lambda \uparrow S} U(\Lambda, S - \Lambda) = 0$$

then by putting $$W(y) = \begin{cases} 1 & \text{if} \quad y = x, \\ 0 & \text{otherwise,} \end{cases}$$

we have $W^+ = \rho^+(\{x\})$, $W^- = \rho^-(\{x\})$. Thus phase transition cannot occur in this case if for some $\epsilon > 0$ either (i) $U(y, S) \geqslant 0$ for all $y \in S$, or (ii) $-U(y, S) \geqslant 0$ for all $y \in S$. However this result is worse than that got by using Propositions 9.3 and 9.4. The most effective use of Theorem 9.3 is when it is known *a priori* that $U(x, S), \rho^+(\{x\})$, and $\rho^-(\{x\})$ are independent of $x \in S$. (This will happen if there is some extra structure on S with respect to which U has enough symmetry or homogeneity properties.) We can then let $W(y) = 1$ for all $y \in S$ and thus Theorem 9.3 gives that if

$$\liminf_{\Lambda \uparrow S} \frac{U(\Lambda, S - \Lambda)}{|\Lambda|} = 0$$

then phase transition cannot occur if V is not an Ising potential.

As an example of this let $S = \mathbb{Z}^\nu$ (for some $\nu \geqslant 1$), where $\mathbb{Z}^\nu$ denotes the points of $\mathbb{R}^\nu$ that have integer coordinates. Let V be a pair potential on $\mathbb{Z}^\nu$ with associated bilinear form U; we will say that V is *translation invariant* if for any $x, y, z \in \mathbb{Z}^\nu$ we $U(x, y) = U(x+z, y+z)$.

THEOREM 9.4 Let $V \in H(\mathbb{Z}^\nu)$ be an attractive translation invariant pair potential. Then phase transition cannot occur if V is not an Ising potential.

Proof Let U be the bilinear form associated with V. We clearly have that $U(x, \mathbb{Z}^\nu)$, $\rho^+(\{x\})$ and $\rho^-(\{x\})$ are independent of $x \in \mathbb{Z}^\nu$, thus by Theorem 9.3 we need only check that

$$\liminf_{\Lambda \uparrow \mathbb{Z}^\nu} \frac{U(\Lambda, \mathbb{Z}^\nu - \Lambda)}{|\Lambda|} = 0.$$

For $n \geqslant 1$ let $a(n) = \sum_{\substack{x \in \mathbb{Z}^\nu \\ n-1<\|x\|\leqslant n}} U(0, x)$.

Then $a(n) \geqslant 0$ and $\sum_{n=1}^{\infty} a(n) = \sum_{y \neq 0} U(0, y) < \infty$.

Let $\Lambda_n \in \mathscr{C}(\mathbb{Z}^\nu)$ be the 'cube' defined by

$$\Lambda_n = \{(x_1, \ldots, x_\nu) \in \mathbb{Z}^\nu : |x_i| \leqslant n, i = 1, \ldots, \nu\}$$

(and put $\Lambda_{-1} = \varnothing$). Then

$$\begin{aligned} U(\Lambda_n, \mathbb{Z}^\nu - \Lambda_n) &= \sum_{r=0}^{n} \sum_{y \in \Lambda_r - \Lambda_{r-1}} U(y, \mathbb{Z}^\nu - \Lambda_n) \\ &\leqslant \sum_{r=0}^{n} |\Lambda_r - \Lambda_{r-1}| \sum_{s=n-r}^{\infty} a(s+1) \\ &= \sum_{s=0}^{\infty} a(s+1) \sum_{r=\max\{n-s,\,0\}}^{n} |\Lambda_r - \Lambda_{r-1}| \\ &= \sum_{s=0}^{\infty} a(s+1)\,(|\Lambda_n| - |\Lambda_{m(n,s)}|), \end{aligned}$$

where $m(n, s) = \max\{n-s, 0\} - 1$. Thus

$$\frac{U(\Lambda_n, \mathbb{Z}^\nu - \Lambda_n)}{|\Lambda_n|} \leqslant \sum_{s=0}^{\infty} a(s+1) \left[1 - \frac{|\Lambda_{m(n,s)}|}{|\Lambda_n|}\right].$$

Now given $\epsilon > 0$ there exists n_0 such that

$$\sum_{s=n_0}^{\infty} a(s+1) \leqslant \tfrac{1}{2}\epsilon.$$

Therefore if $n > n_0$ we have

$$\begin{aligned} \sum_{s=0}^{\infty} a(s+1) \left[1 - \frac{|\Lambda_{m(n,s)}|}{|\Lambda_n|}\right] &\leqslant \tfrac{1}{2}\epsilon + \sum_{s=0}^{n_0-1} a(s+1) \left[1 - \frac{|\Lambda_{m(n,s)}|}{|\Lambda_n|}\right] \\ &\leqslant \tfrac{1}{2}\epsilon + \left[1 - \frac{|\Lambda_{n-n_0}|}{|\Lambda_n|}\right] \sum_{s=0}^{n_0-1} a(s+1). \end{aligned}$$

But
$$\frac{|\Lambda_{n-n_0}|}{|\Lambda_n|} = \left[\frac{2(n-n_0)+1}{2n+1}\right]^\nu,$$

and thus if n is large enough we have

$$\left[1 - \frac{|\Lambda_{n-n_0}|}{|\Lambda_n|}\right] \sum_{s=0}^{n_0-1} a(s+1) \leqslant \tfrac{1}{2}\epsilon$$

and hence
$$\frac{U(\Lambda_n, \mathbb{Z}^\nu - \Lambda_n)}{|\Lambda_n|} < \epsilon.$$

Therefore
$$\liminf_{\Lambda \uparrow \mathbb{Z}^\nu} \frac{U(\Lambda, \mathbb{Z}^\nu - \Lambda)}{|\Lambda|} = 0. \square$$

NOTES The proof of Theorem 9.1 given here comes from Ruelle (1969); Theorem 9.1 is a generalization of the circle theorem of Yang and Lee (1952). The technique used to prove Theorem 9.3 is due to Lebowitz and Martin-Löf (1972) who used it to prove Theorem 9.4, a result which was also proved by Ruelle (1971b).

10. *Examples of phase transition*

At this point the reader may well be wondering whether phase transition occurs at all. In this chapter we will hopefully remove any such doubts from the reader's mind by giving examples of potentials for which phase transition does occur. From the results of the previous chapters it would seem that if phase transition does occur then it is most likely to occur for Ising potentials. For most of this chapter we will thus be concerned with Ising potentials.

PROPOSITION 10.1 Let V be a supermodular Ising potential. Then phase transition occurs for V if and only if there exists $x \in S$ and $\alpha > 1$ such that for all $\Lambda \in \mathscr{C}(S)$ with $x \in \Lambda$ we have

$$[\sum_{x \notin B \subset \Lambda} \exp V(B)]/[\sum_{x \in A \subset \Lambda} \exp V(A)] \geqslant \alpha.$$

Proof We have

$$\begin{aligned}[\sum_{x \notin B \subset \Lambda} \exp V(B)]/[\sum_{x \in A \subset \Lambda} \exp V(A)] &= [1-\rho_\Lambda^-(\{x\})]/\rho_\Lambda^-(\{x\}) \\ &= \rho_\Lambda^+(\{x\})/\rho_\Lambda^-(\{x\})\end{aligned}$$

(by Proposition 8.9). Note that we have

$$\rho_\Lambda^+(\{x\})/\rho_\Lambda^-(\{x\}) \downarrow \rho^+(\{x\})/\rho^-(\{x\}) \quad \text{as} \quad \Lambda \uparrow S.$$

The result thus follows by Theorem 8.1. $\square$

It will now be convenient to assume that S has more structure and we will suppose that the points of S are the vertices of a graph $\mathscr{G} = (S, e)$, where e is the set of edges of $\mathscr{G}$. As in Chapter 4 we do not allow $\mathscr{G}$ to have any multiple edges or loops. We will say that $x, y \in S$ are *neighbours* if there is an edge of the graph between them (and we will denote this edge by $\langle x, y \rangle$);

we will assume that each $x \in S$ has only a finite number of neighbours. If $A \subset S$ and $z_1, \ldots, z_n \in A$ then $z_1, \ldots, z_n$ is called a *path from x to y in A* if $x = z_1, y = z_n$ and if z_i and z_{i+1} are neighbours for $i = 1, \ldots, n-1$; $A \subset S$ is called *connected* if given $x \neq y \in A$ then there exists a path from x to y in A. From now on we will assume that S is connected. If α is a finite subset of e and $x \in S$ then we define $I(x, \alpha) \subset S$ by

$I(x, \alpha) = \{x\} \cup \{y \in S$: there exists a path $x = z_1, \ldots, z_n = y$ in S such that $\langle z_i, z_{i+1}\rangle \notin \alpha, i = 1, \ldots, n-1\}$.

Let $x \in \Lambda \in \mathscr{C}(S)$ and let α be a finite subset of e; we will say that *α blocks x in Λ* if $I(x, \alpha) \subset \Lambda$. If α blocks x in Λ but no proper subset of α blocks x in Λ then we will call α a *border round x in Λ*; $|\alpha|$ is called the *length* of the border.

An element Λ of $\mathscr{C}(S)$ is called a *simplex* if given any $x, y \in \Lambda$ with $x \neq y$ then x and y are neighbours. A potential $V \in H(S)$ will be called *nearest neighbour* if $J_V(A) = 0$ whenever A is not a simplex. Let us consider only pair potentials, thus let V be a nearest neighbour supermodular Ising pair potential with associated bilinear form U. Therefore we have $U(x,y) \geqslant 0$ if $x \neq y$ and $U(x,y) = 0$ if x and y are not neighbours; also $U(x, S) = 0$ for all $x \in S$.

Proposition 10.2 Suppose that $U(x,y) \geqslant b > 0$ whenever x and y are neighbours. Let $x \in \Lambda \in \mathscr{C}(S)$; then

$$\rho_\Lambda^-(\{x\}) \leqslant \sum_{n=1}^{\infty} r_n(x, \Lambda) \exp(-nb),$$

where $r_n(x, \Lambda)$ is the number of borders round x in Λ with length n.

Proof Note that if α is a border round x in Λ and $a \in \alpha$ with $a = \langle z_1, z_2\rangle$ then at least one of z_1, z_2 is in Λ. If $A \subset \Lambda$ and α is a border round x in Λ then we will say that A *contains* α if given any $a \in \alpha$ with $a = \langle z_1, z_2\rangle$ then exactly one of z_1, z_2 is in A. We have the following estimate:

Lemma 10.1 Let α be a border round x in Λ and let $B_\alpha = \{A \subset \Lambda : A \text{ contains } \alpha\}$. Then

$$\sum_{A \in B_\alpha} \pi_\Lambda^-(A) \leqslant \exp(-|\alpha| b).$$

Proof Let $W: e \to \mathbb{R}$ be given by

$$W(\langle z_1, z_2 \rangle) = -U(z_1, z_2).$$

Then if $A \in \mathscr{C}(S)$ we have

$$V(A) = U(A, A) = U(A, S) - U(A, S-A) = -U(A, S-A)$$
$$= \sum_{a \in e} A(a)\, W(a),$$

where

$$A(\langle z_1, z_2 \rangle) = \begin{cases} 1 & \text{if exactly one of } z_1, z_2 \text{ is in } A, \\ 0 & \text{otherwise.} \end{cases}$$

Now define $g: \mathscr{P}(\Lambda) \to \mathscr{P}(\Lambda)$ by

$$g(A) = [A \cap (\Lambda - I(x, \alpha))] \cup [I(x, \alpha) - A]$$

(i.e. 0's and 1's are interchanged on $I(x, \alpha)$; thus in the notation of Chapter 7 we have $g = \tau_R$, with $R = I(x, \alpha)$); g is clearly a bijection from $\mathscr{P}(\Lambda)$ to $\mathscr{P}(\Lambda)$. If $A \in B_\alpha$ then

$$V(g(A)) = \sum_{a \in e} [g(A)](a)\, W(a)$$
$$= \sum_{a \in e - \alpha} [g(A)](a)\, W(a) = \sum_{a \in e - \alpha} A(a)\, W(a)$$
$$= V(a) - \sum_{a \in \alpha} W(a) \geqslant V(a) + |\alpha|\, b;$$

i.e.

$$V(A) \leqslant V(g(A)) - |\alpha|\, b.$$

Therefore

$$\sum_{A \in B_\alpha} \pi_\Lambda^-(A) = [\sum_{A \in B_\alpha} \exp V(A)] / [\sum_{A \subset \Lambda} \exp V(A)]$$
$$\leqslant [\sum_{A \in B_\alpha} \exp V(A)] / [\sum_{A \in B_\alpha} \exp V(g(A))]$$
$$\leqslant \exp(-|\alpha|\, b). \square$$

We now continue with the proof of Proposition 10.2 by showing:

LEMMA 10.2 Let $A \subset \Lambda$ with $x \in A$. Then there exists a border α round x in Λ such that A contains α.

Proof Let $\beta = \{a \in e : A(a) = 1\}$; then clearly β blocks x in Λ. There thus exists a border α round x in Λ with $\alpha \subset \beta$ and by construction A contains α. □

The proof of Proposition 10.2 is now easily completed. Let $D(n)$ denote the set of borders round x in Λ of length n and let $D = \bigcup_{n=1}^{\infty} D(n)$. Then by Lemma 10.2 we have

$$\rho_\Lambda^-(\{x\}) \leqslant \sum_{\alpha \in D} \Big[\sum_{A \in B_\alpha} \pi_\Lambda^-(A) \Big]$$

and therefore using Lemma 10.1 we get

$$\begin{aligned} \rho_\Lambda^-(\{x\}) &\leqslant \sum_{n=1}^{\infty} \sum_{\alpha \in D(n)} \sum_{A \in B_\alpha} \pi_\Lambda^-(A) \\ &\leqslant \sum_{n=1}^{\infty} \sum_{\alpha \in D(n)} \exp(-nb) \\ &= \sum_{n=1}^{\infty} r_n(x, \Lambda) \exp(-nb). \square \end{aligned}$$

Note that if α is a border round x in Λ and $\Lambda' \supset \Lambda$ then α is a border round x in Λ'; let us call α a border round x if it is a border round x in Λ for some $\Lambda \in \mathscr{C}(S)$, and let $r_n(x)$ denote the number of borders round x having length n. It is clear that $r_n(x, \Lambda) \leqslant r_n(x)$ for all $\Lambda \in \mathscr{C}(S)$. Thus from Proposition 10.2 we immediately get:

THEOREM 10.1 Let V be a nearest neighbour supermodular Ising pair potential with associated bilinear form U, and suppose that $U(x, y) \geqslant b > 0$ whenever x and y are neighbours. Suppose also that

$$\sum_{n=1}^{\infty} r_n(x) \exp(-nb) < \tfrac{1}{2}$$

for some $x \in S$. Then phase transition occurs for V.

Let us call a graph $\mathscr{G} = (S, e)$ a *phase transition graph* if given any nearest neighbour supermodular Ising pair potential

$V \in H(S)$ with associated bilinear form U and such that for some $\epsilon > 0$ $U(x, y) \geqslant \epsilon$ whenever x and y are neighbours then phase transition occurs for tV for some $t > 0$. (Note that by Theorem 8.4, if phase transition occurs for tV then phase transition also occurs for sV for all $s > t$.)

Proposition 10.3 Let $\mathscr{G} = (S, e)$ and suppose there exists $x \in S$ and $K > 0$ such that $r_n(x) \leqslant \exp(Kn)$ for all $n \geqslant 1$. Then $\mathscr{G}$ is a phase transition graph.

Proof This follows from Theorem 10.1 since if t is large enough then

$$\sum_{n=1}^{\infty} \exp((K - \epsilon t)\, n) < \tfrac{1}{2}. \square$$

For $\nu \in \mathbb{N}$ let $\mathbb{Z}^\nu$ denote the points of $\mathbb{R}^\nu$ that have integer coordinates, considered as a graph in the usual way (i.e. $x, y \in \mathbb{Z}^\nu$ are neighbours if $\|x - y\| = 1$). $\mathbb{Z}^1$ certainly does not satisfy the hypothesis of Proposition 10.3 (since $r_2(x) = \infty$), but for $\nu \geqslant 2$ it is possible to check that $\mathbb{Z}^\nu$ does. However to do this for $\nu > 2$ is very involved, so to show that $\mathbb{Z}^\nu$, for $\nu \geqslant 2$, is a phase transition graph we proceed as follows: If $\mathscr{G} = \{S, e\}$ and $\mathscr{G}' = \{S', e'\}$ are graphs then $\mathscr{G}'$ will be called a subgraph of $\mathscr{G}$ if $S' \subset S$ and $e' \subset e$ (i.e. if $x, y \in S'$ and there is an edge between x and y in $\mathscr{G}'$ then there is an edge between x and y in $\mathscr{G}$).

Proposition 10.4 If $\mathscr{G}'$ is a subgraph of $\mathscr{G}$ and $\mathscr{G}'$ is a phase transition graph then $\mathscr{G}$ is a phase transition graph.

Proof Let $\mathscr{G} = (S, e)$, $\mathscr{G}' = (S', e')$ and let $V \in H(S)$ be a nearest neighbour supermodular Ising pair potential with associated bilinear form U, and suppose that for some $\epsilon > 0$ we have $U(x, y) \geqslant \epsilon$ whenever x and y are neighbours in $\mathscr{G}$. Define $V' \in H(S')$ by letting its associated bilinear form U' be given by

$$\begin{aligned} U'(x,y) &= U(x,y) \quad \text{if} \quad x, y \in S' \text{ are neighbours in } \mathscr{G}', \\ &= 0 \qquad\qquad \text{for all other } x, y \in S' \quad \text{with} \quad x \neq y, \\ U'(x,x) &= -\sum_{x \neq y \in S'} U'(x,y). \end{aligned}$$

Thus $U'(x, y) \geqslant \epsilon$ whenever x and y are neighbours in $\mathscr{G}'$ and since V' is a nearest neighbour supermodular Ising pair potential on S' we have there exists $t_0 > 0$ such that phase transition occurs for $t_0 V'$. Now define $V^* \in H(S)$ by letting its associated bilinear form U^* be given by

$$U^*(x, y) = U'(x, y) \quad \text{if} \quad x, y \in S',$$
$$= 0 \quad \text{otherwise.}$$

Then V^* is a nearest neighbour supermodular Ising pair potential on S, and we leave it to the reader to verify that if $t > 0$ then phase transition occurs for tV^* if and only if it occurs for tV'. Thus phase transition occurs for $t_0 V^*$. But it is easy to check that we have $I_{t_0 V}(A) \geqslant I_{t_0 V^*}(A)$ for all $A \in \mathscr{C}(S)$ and therefore by Theorem 8.4 we have that phase transition occurs for $t_0 V$. Hence $\mathscr{G}$ is a phase transition graph. □

We can clearly regard $\mathbb{Z}^2$ as a subgraph of $\mathbb{Z}^\nu$ for any $\nu \geqslant 2$, thus to show that $\mathbb{Z}^\nu$ is a phase transition graph for $\nu \geqslant 2$ we need only show that $\mathbb{Z}^2$ is a phase transition graph.

PROPOSITION 10.5 $\mathbb{Z}^2$ is a phase transition graph.

Proof We will show that $\mathbb{Z}^2$ satisfies the hypotheses of Proposition 10.3. Firstly we define a graph $\mathscr{E}$ whose vertices are the edges of $\mathbb{Z}^2$; thus it is natural to consider the vertices of $\mathscr{E}$ to be the mid-points of the edges of $\mathbb{Z}^2$, i.e. points of $\mathbb{R}^2$ that are either of the form $(n, m + \frac{1}{2})$ or of the form $(n + \frac{1}{2}, m)$, with $(n, m) \in \mathbb{Z}^2$. We define the neighbours of (x, y) in $\mathscr{E}$ to be $(x - \frac{1}{2}, y - \frac{1}{2})$, $(x - \frac{1}{2}, y + \frac{1}{2})$, $(x + \frac{1}{2}, y - \frac{1}{2})$ and $(x + \frac{1}{2}, y + \frac{1}{2})$; thus $\mathscr{E}$ is isomorphic to $\mathbb{Z}^2$ (see Fig. 1). We leave it to the reader to check that the borders round 0 in $\mathbb{Z}^2$ correspond exactly to the circuits in $\mathscr{E}$ round 0 (where a circuit in $\mathscr{E}$ is a path $z_0, z_1, \ldots, z_n$ with $z_0 = z_n$ and $z_1, \ldots, z_n$ distinct). Now the number of circuits in $\mathscr{E}$ of length n and which include a given point $w \in \mathscr{E}$ must be less than 3^n (since given $w = z_0, z_1, \ldots, z_k$ as part of a circuit, then there are at most 3 choices for z_{k+1}). If a circuit has length n and goes round 0 then each point of the circuit must be within $\frac{1}{2}n$ steps from 0; therefore the number of circuits in $\mathscr{E}$ of length n which go round 0 is certainly less than $n^2 3^n$. Therefore by Proposition 10.3 we have that $\mathbb{Z}^2$ is a phase transition graph. □

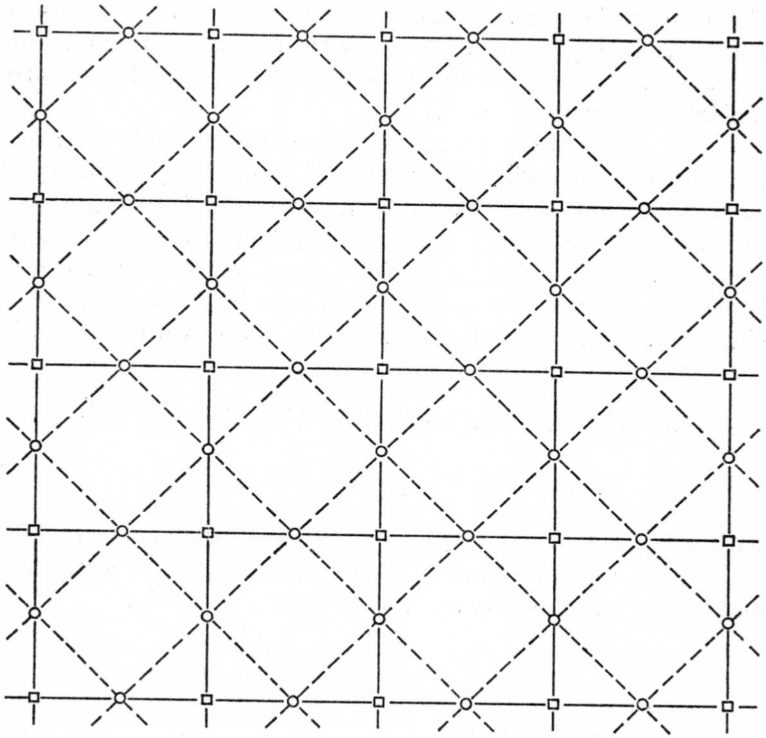

□ vertex of $\mathbb{Z}^2$ ○ vertex of $\mathscr{E}$
——— edge of $\mathbb{Z}^2$ - - - - edge of $\mathscr{E}$

Fig. 1

To complete the picture for $\mathbb{Z}^\nu$ we should now show that $\mathbb{Z}$ $(= \mathbb{Z}^1)$ is not a phase transition graph. However, for the moment we will leave this as an exercise for the reader, though a proof does appear at the end of this chapter. We now turn our attention to a very simple class of graphs for which explicit calculations can be made. A graph $\mathscr{G} = (S, e)$ will be called a *tree* if it is connected and contains no circuits. Thus $\mathscr{G}$ is a tree if and only if given $x \neq y \in S$ then there exists a unique path $x = z_1, \ldots, z_n = y$ from x to y with $z_1, \ldots, z_n$ distinct. For $n \geqslant 2$ let $\mathscr{T}^n$ denote the tree in which every vertex has exactly n neighbours; thus $\mathscr{T}^2 = \mathbb{Z}$.

Let $\mathscr{G} = (S, e)$ be a tree; the simple structure of $\mathscr{G}$ allows us to define elements of $\mathscr{S}(S)$ by actually giving the probability of each finite dimensional cylinder. We need the following definitions: let $x \in S$ and let $y_1, y_2 \in S$ be neighbours; we will say y_1 is closer to x than y_2 if any path from x to y_2 contains y_1; since $\mathscr{G}$ is a tree we must have either y_1 is closer to x than y_2 or y_2 is closer to x than y_1 (and exactly one of these two happens). If $\Lambda \in \mathscr{C}(S)$ with Λ connected then let

$$e_\Lambda = \{\langle z_1, z_2\rangle \in e : z_1, z_2 \text{ are neighbours and } z_1, z_2 \in \Lambda\}.$$

Let
$$Q = \begin{pmatrix} p & 1-p \\ 1-q & q \end{pmatrix},$$

with $0 < p < 1$, $0 < q < 1$, be a 2×2 stochastic matrix and let $r(0), r(1) \geqslant 0$ with $r(0) + r(1) = 1$; let $\Lambda \in \mathscr{C}(S)$ with Λ connected and take $A \subset \Lambda$. For any $x \in \Lambda$ we define $h_x : e_\Lambda \to \mathbb{R}$ by

$$h_x(\langle z_1, z_2\rangle) = \begin{cases} p & \text{if } z_1 \notin A, z_2 \notin A, \\ q & \text{if } z_1 \in A, z_2 \in A, \\ 1-p & \text{if } z_1 \notin A, z_2 \in A, \\ & \quad \text{and } z_1 \text{ is closer to } x \text{ than } z_2, \\ 1-q & \text{if } z_1 \in A, z_2 \notin A, \\ & \quad \text{and } z_1 \text{ is closer to } x \text{ than } z_2, \end{cases}$$

and let $g : \Lambda \to \mathbb{R}$ be given by

$$g(x) = \begin{cases} r(1) \prod_{a \in e_\Lambda} h_x(a) & \text{if } x \in A, \\ r(0) \prod_{a \in e_\Lambda} h_x(a) & \text{if } x \notin A. \end{cases}$$

Lemma 10.3 $g(x)$ is independent of $x \in \Lambda$ if and only if

$$r(0)(1-p) = r(1)(1-q).$$

Proof This is easily checked. □

We will suppose now that we do have $r(0)\,(1-p) = r(1)\,(1-q)$, and we can thus define

$$\mu([A, \Lambda]) = g(x) \quad \text{(for any } x \in \Lambda\text{)},$$

whenever $A \subset \Lambda \in \mathscr{C}(S)$ and Λ is connected. It is not difficult to see that this uniquely defines an element $\mu \in \mathscr{S}(S)$; and of course, given any $A \subset \Lambda \in \mathscr{C}(S)$ we can explicitly compute the value of $\mu([A, \Lambda])$. The above construction of μ may seem somewhat involved but it can be checked that μ is just defined to make it act like a Markov chain along any path in S.

PROPOSITION 10.6 $\mu \in \mathscr{G}_V$, where $V \in H(S)$ is the nearest neighbour pair potential with associated bilinear form U given by

$$U(x,y) = \begin{cases} \log\left[\left(\dfrac{1-q}{p}\right)^{m(x)}\left(\dfrac{1-p}{1-q}\right)\right] & \text{if } x = y, \\ \tfrac{1}{2}\log\left[\dfrac{pq}{(1-p)\,(1-q)}\right] & \text{if } x \text{ and } y \text{ are neighbours}, \\ 0 \quad \text{otherwise}, \end{cases}$$

where $m(x)$ is the number of neighbours of x.

Proof We leave this as an exercise for the reader. □

We will now restrict ourselves to the case when $\mathscr{G} = \mathscr{T}^m$ for some $m \geqslant 2$ (and thus $m(x) = m$ for all $x \in S$), and we consider the converse of the above construction. That is, given $v_0, v_1 \in \mathbb{R}$ and defining a pair potential $V \in H(S)$ by letting its associated bilinear form U be given by

$$U(x,y) = \begin{cases} v_0 & \text{if } x = y, \\ v_1 & \text{if } x \text{ and } y \text{ are neighbours}, \\ 0 & \text{otherwise}, \end{cases}$$

then can p, q and r be chosen so that the above construction gives us an element of $\mathscr{G}_V$? Now since we have $r(0)+r(1) = 1$ and $r(0)\,(1-p) = r(1)\,(1-q)$ it is clear that r is determined by p and q. By Proposition 10.6 we must choose p, q so that

$$\left(\frac{1-q}{p}\right)^m \left(\frac{1-p}{1-q}\right) = \exp v_0$$

and
$$\frac{pq}{(1-p)(1-q)} = \exp 2v_1.$$

Therefore, putting $a = \exp v_0, b = \exp 2v_1$ (and thus $a > 0, b > 0$) and making the substitution

$$\alpha = \frac{p}{1-p}, \quad \beta = \frac{q}{1-q},$$

the problem becomes: do there exist $\alpha > 0, \beta > 0$ so that $\alpha\beta = b$ and

$$\left(\frac{1+\alpha}{1+\beta}\right)^{m-1} \frac{1}{\alpha^m} = a?$$

This, of course, reduces to: does there exist $\alpha > 0$ such that

$$\left(\frac{1+\alpha}{b+\alpha}\right)^{m-1} \frac{1}{\alpha} = a?$$

The answer is easily seen to be yes, since if we define $f: \mathbb{R}^+ \to \mathbb{R}^+$ by

$$f(x) = \left(\frac{1+x}{b+x}\right)^{m-1},$$

then f is bounded and thus the curve $y = f(x)$ must intersect the line $y = ax$. Therefore our construction gives an element of $\mathcal{G}_V$ for any $v_0, v_1 \in \mathbb{R}$. Note that if the equation

$$\left(\frac{1+x}{b+x}\right)^{m-1} = ax$$

has more than one solution then our construction gives more than one element of $\mathcal{G}_V$ for the corresponding $v_0 = \log a$, $v_1 = \frac{1}{2}\log b$; and thus we would have explicitly exhibited phase transition. It is thus worthwhile to examine how many solutions the equation $f(x) = ax$ has.

PROPOSITION 10.7 The equation

$$\left(\frac{1+x}{b+x}\right)^{m-1} = ax$$

(with $x \geqslant 0, m \geqslant 2, a > 0, b > 0$) has one solution if either $m = 2$

or $b \leqslant (m/m-2)^2$. If $m > 2$ and $b > (m/m-2)^2$ then there exist $\eta_1(b,m), \eta_2(b,m)$ with $0 < \eta_1(b,m) < \eta_2(b,m)$ such that the equation has three solutions if $\eta_1(b,m) < a < \eta_2(b,m)$ and has two solutions if either $a = \eta_1(b,m)$ or $a = \eta_2(b,m)$. In fact

$$\eta_i(b,m) = \frac{1}{x_i}\left(\frac{1+x_i}{b+x_i}\right)^{m-1},$$

where x_1, x_2 are the solutions of

$$x^2+[2-(b-1)(m-2)]\,x+b = 0.$$

Proof As before let

$$f(x) = \left(\frac{1+x}{b+x}\right)^{m-1};$$

thus we have $\quad f'(x) = (m-1)(b-1)\dfrac{(1+x)^{m-2}}{(b+x)^m},$

$$f''(x) = (m-1)(b-1)(b(m-2)-m-2x)\frac{(1+x)^{m-3}}{(b+x)^{m+1}}.$$

In particular if $b \leqslant 1$ then f is decreasing and there can only be one solution of $f(x) = ax$; thus we can restrict ourselves to the case $b > 1$. If $m = 2$ then $f'' \leqslant 0$ and thus f is concave increasing; hence there is only one solution when $m = 2$. For $m > 2$ we have that f is convex for $x < \frac{1}{2}[b(m-2)-m]$ and is concave for $x > \frac{1}{2}[b(m-2)-m]$; thus there are at most three solutions to $f(x) = ax$. In fact it is quite easy to see that there is more than one solution if and only if there is more than one solution to $xf'(x) = f(x)$, which is the same as

$$x^2+[2-(b-1)(m-2)]\,x+b = 0.$$

With the help of a little elementary analysis the proof is readily completed. □

Note that we can only construct explicit examples of phase transition when $b > 1$, which means $v_1 > 0$, i.e. that the potential is supermodular. However for supermodular potentials we will now show that our construction exhibits phase transition whenever it can occur.

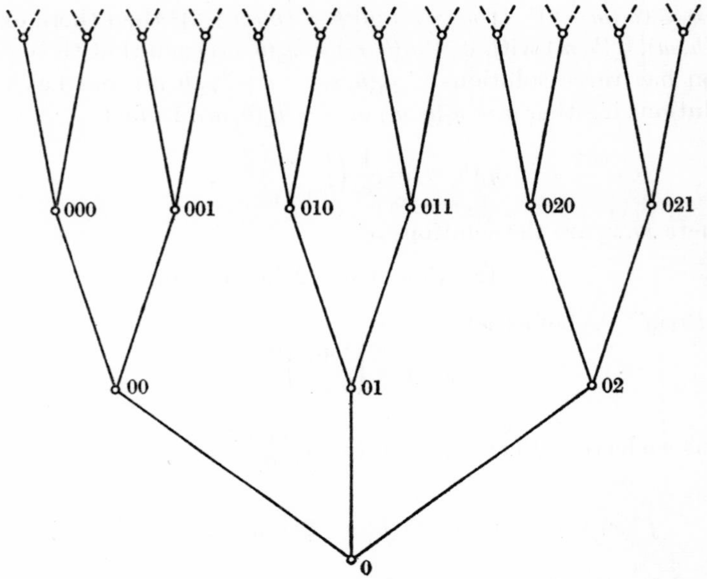

Fig. 2

Let V be a pair potential on $\mathscr{T}^m$ defined in terms of v_0, v_1 as before, and suppose that $v_1 > 0$. We will see whether phase transition occurs for V by examining the high and low density states for V. It will be convenient to represent $\mathscr{T}^m = (S, e)$ as follows:

$$S = \{\epsilon_1\epsilon_2\ldots\epsilon_n : \epsilon_1 = 0, \epsilon_2 = 0, 1, \ldots, m-1, \quad \text{and}$$
$$\epsilon_\nu = 0, 1, \ldots, m-2 \quad \text{for} \quad \nu \geqslant 3; n = 1, 2, \ldots\};$$

e is defined so that the neighbours of $\epsilon_1\epsilon_2\ldots\epsilon_n$ are exactly $\epsilon_1\epsilon_2\ldots\epsilon_n 0, \ldots, \epsilon_1\ldots\epsilon_n(m-2)$, and $\epsilon_1\ldots\epsilon_{n-1}$ if $n \geqslant 2$, and the neighbours of 0 are $00, 01, \ldots, 0(m-1)$. (See Fig. 2 for the case $m = 3$.) Let $\mathscr{T}_0^m = (S_0, e)$ be the subgraph of $\mathscr{T}^m$ with

$$S_0 = \{\epsilon_1, \ldots\epsilon_n : \epsilon_2 \neq m-1\}.$$

(See Fig. 3, again for the case $m = 3$.) For $N \geqslant 1$ let

$$\Lambda_N = \{\epsilon_1\ldots\epsilon_n : n \leqslant N\}$$

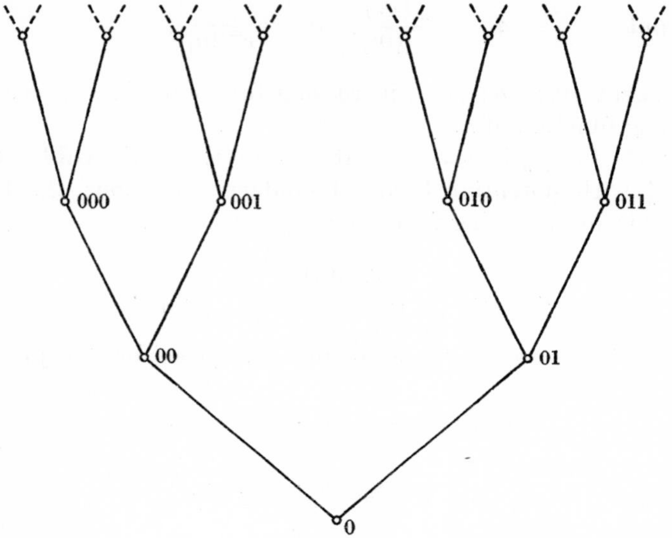

Fig. 3

and let $\Lambda'_N = \Lambda_N \cap S_0$. Let π_n^+ (resp. π_n^-) denote the Gibbs state on Λ_n with potential V and boundary conditions $S-\Lambda_n$ (resp. $\varnothing$); let ρ_n^+ (resp. ρ_n^-) be the correlation function of π_n^+(resp. π_n^-); let π^+ (resp. π^-) denote the high density (resp. low density) state for V, and let ρ^+ (resp. ρ^-) be the correlation function of π^+ (resp. π^-). Then since $\rho^+(\{0\}) = \rho^+(\{x\})$, $\rho^-(\{0\}) = \rho^-(\{x\})$ for all $x \in S$ we have that phase transition occurs for V if and only if $\rho^+(\{0\}) \neq \rho^-(\{0\})$. We also know that

$$\rho_n^+(\{0\}) \downarrow \rho^+(\{0\}), \quad \rho_n^-(\{0\}) \uparrow \rho^-(\{0\}).$$

Define $r_n^-(\{0\}), r_n^-(\varnothing), r_n^+(\{0\}), r_n^+(\varnothing)$ by

$$r_n^-(\{0\}) = \sum_{0 \in A \subset \Lambda'_n} \exp V(A), \quad r_n^-(\varnothing) = \sum_{0 \notin A \subset \Lambda'_n} \exp V(A),$$

$$r_n^+(\{0\}) = \sum_{0 \in A \subset \Lambda'_n} \exp [V(A) + 2U(A, S_0 - \Lambda'_n)],$$

$$r_n^+(\varnothing) = \sum_{0 \notin A \subset \Lambda'_n} \exp [V(A) + 2U(A, S_0 - \Lambda'_n)],$$

and put $$a_n^- = \frac{r_n^-(\varnothing)}{r_n^-(\{0\})}, \quad a_n^+ = \frac{r_n^+(\varnothing)}{r_n^+(\{0\})}.$$

LEMMA 10.4 a_n^+ is an increasing function of n; a_n^- is a decreasing function of n.

Proof Let $\bar{\rho}_n$ be the correlation function of the Gibbs state on Λ_n' with potential V and boundary conditions $\varnothing$. Then $\bar{\rho}_n(\{0\})$ is an increasing function of n, but we have

$$\bar{\rho}_n(\{0\}) = \frac{r_n^-(\{0\})}{r_n^-(\{0\})+r_n^-(\varnothing)} = \frac{1}{1+a_n^-}.$$

Thus a_n^- is a decreasing function of n. The proof for a_n^+ is just the same. □

Let $$a^+ = \lim_{n\to\infty} a_n^+, \quad a^- = \lim_{n\to\infty} a_n^-,$$

and let $$a = \exp v_0, \quad b = \exp 2v_1.$$

PROPOSITION 10.8

(i) $$\rho_n^+(\{0\}) = \left[1+\frac{1}{a}\left(\frac{1+a_{n-1}^+}{b+a_{n-1}^+}\right)^m\right]^{-1},$$

$$\rho_n^-(\{0\}) = \left[1+\frac{1}{a}\left(\frac{1+a_{n-1}^-}{b+a_{n-1}^-}\right)^m\right]^{-1},$$

(ii) phase transition occurs for V if and only if $a^+ \neq a^-$;

(iii) $$a_n^+ = \frac{1}{a}\left(\frac{1+a_{n-1}^+}{b+a_{n-1}^+}\right)^{m-1}, \quad a_n^- = \frac{1}{a}\left(\frac{1+a_{n-1}^-}{b+a_{n-1}^-}\right)^{m-1}.$$

Proof This is straightforward and is left to the reader. □

From Proposition 10.8 it follows that a^+ and a^- are non-negative solutions of

$$x = \frac{1}{a}\left(\frac{1+x}{b+x}\right)^{m-1},$$

and in fact a^+ (resp. a^-) must be the smallest (resp. largest) non-negative solution. We have thus shown:

Proposition 10.9 Phase transition occurs for V (with $v_1 > 0$) if and only if the equation

$$\left(\frac{1+x}{b+x}\right)^{m-1} = ax$$

has more than one non-negative solution.

This is of course the same equation that appeared before. As noted in Proposition 10.7 the equation has only one solution if $m = 2$ and hence we have:

Proposition 10.10 $\mathbb{Z}\,(=\mathscr{T}^2)$ is not a phase transition graph.

Using Proposition 10.9 and Theorem 8.4 it is not hard to see that, for $m \geqslant 3$, $\mathscr{T}^m$ is a phase transition graph. It is also worth pointing out that if $m \geqslant 3$ then V need not be an Ising potential on $\mathscr{T}^m$ for phase transition to occur. This shows the necessity of some condition like

$$\liminf_{\Lambda \uparrow S} \frac{U(\Lambda, S-\Lambda)}{W(\Lambda)} = 0$$

to make Theorem 9.3 work.

Notes The basic idea used to prove Proposition 10.2 and Proposition 10.5 is due to Peierls (1936). Peierls' method was made rigorous for $\mathbb{Z}^2$ and $\mathbb{Z}^3$ by Griffiths (1964) and Dobrushin (1965). The graphs looked at in the latter half of the chapter were suggested to the author by Frank Spitzer.

11. *The extreme points of* $\mathscr{G}_V$

Let S be a countable set (we will not assume that S has any extra structure in this chapter), and let $V \in H(S)$. By Proposition 5.2 we know that $\mathscr{G}_V$, the set of Gibbs states with potential V, is a non-empty, compact convex subset of $\mathscr{S}(S)$. In this chapter we will characterize the extreme points of $\mathscr{G}_V$ (where $\mu \in \mathscr{G}_V$ is an extreme point of $\mathscr{G}_V$ if there does not exist $\nu_1, \nu_2 \in \mathscr{G}_V$ with $\nu_1 \neq \nu_2$ and $\mu = \frac{1}{2}(\nu_1 + \nu_2)$).

We will need the following notation and definitions: If $A \in \mathscr{P}(S)$ then as before $\mathscr{F}(A)$ denotes the Borel subsets of the compact Hausdorff space $\mathscr{P}(A)$; we let $\overline{\mathscr{F}}(A)$ denote the sub-σ-algebra of $\mathscr{F}(S)$ generated by

$$\{[B, \Lambda] : B \subset \Lambda \in \mathscr{C}(A)\}.$$

Thus $\overline{\mathscr{F}}(A)$ consists of those elements of $\mathscr{F}(S)$ that are of the form

$$\{X \cup B : X \in \Omega, B \subset S - A\} \quad \text{for some} \quad \Omega \in \mathscr{F}(A).$$

If Y is a set and $\mathscr{B}$ a σ-algebra of subsets of Y then we let $M(Y, \mathscr{B})$ denote the real vector space of functions from Y to $\mathbb{R}$ that are $\mathscr{B}$-measurable; if μ is a measure on $(Y, \mathscr{B})$ then $M(Y, \mathscr{B}, \mu)$ will denote the quotient space of $M(Y, \mathscr{B})$ with the subspace of functions that are zero μ-a.e. Let $\mathscr{B}'$ be a sub σ-algebra of $\mathscr{B}$, then we let $\mathscr{B}'_\mu$ denote the smallest σ-algebra containing $\mathscr{B}'$ and $\{B \in \mathscr{B} : \mu(B) = 0\}$. It is well known that $\mathscr{B}'_\mu$ consists exactly of those subsets of Y of the form BN, with $B \in \mathscr{B}'$ and $N \in \mathscr{B}$ with $\mu(N) = 0$ (where as before BN denotes the symmetric difference of B and N, i.e. $BN = (B - N) \cup (N - B)$). Note that if

$$f \in M(Y, \mathscr{B}'_\mu)$$

then there exists $f' \in M(Y, \mathscr{B}')$ with $f' = f$ μ-a.e. We will regard $M(Y, \mathscr{B}')$ as a subspace of $M(Y, \mathscr{B})$ and also regard $M(Y, \mathscr{B}'_\mu, \mu)$

as a subspace of $M(Y, \mathscr{B}, \mu)$. For $1 \leqslant p \leqslant \infty$ we let $L^p(Y, \mathscr{B}, \mu)$ denote the vector subspace of $M(Y, \mathscr{B}, \mu)$ of those functions whose pth powers are integrable (or are bounded if $p = \infty$), considered as a Banach space in the usual way. If $\mathscr{B}'$ is a sub σ-algebra of $\mathscr{B}$ then $L^p(Y, \mathscr{B}'_\mu, \mu)$ is a closed subspace of

$$L^p(Y, \mathscr{B}, \mu).$$

Let $\mu \in \mathscr{S}(S)$ and $A \in \mathscr{P}(S)$; we will write $\overline{\mathscr{F}}_\mu(A)$ instead of $(\overline{\mathscr{F}}(A))_\mu$. For $1 \leqslant p \leqslant \infty$, $\mu \in \mathscr{S}(S)$ and $A \in \mathscr{P}(S)$ we let

$$L^p_\mu(A) = L^p(\mathscr{P}(S), \overline{\mathscr{F}}_\mu(A), \mu);$$

we will write L^p_μ instead of $L^p_\mu(S)$.

PROPOSITION 11.1 Let $\mu \in \mathscr{S}(S)$ and $A \in \mathscr{P}(S)$. Then there is an isometry

$$i_A : L^p(\mathscr{P}(A), \mathscr{F}(A), r_A(\mu)) \to L^p_\mu(A)$$

given by defining

$$i_A(f)(X) = f(X \cap A) \quad \text{for} \quad X \in \mathscr{P}(S)$$

and $$f \in L^p(\mathscr{P}(A), \mathscr{F}(A), r_A(\mu)).$$

Proof It is easy to check that i_A is well-defined and has the right properties. □

For $1 \leqslant p \leqslant \infty$ let $I^p_\mu = \bigcap_{\Lambda \in \mathscr{C}(S)} L^p_\mu(S - \Lambda)$;

thus I^p_μ is a closed subspace of L^p_μ which contains at least the constant functions. Let

$$\overline{\mathscr{F}}^0_\mu = \bigcap_{\Lambda \in \mathscr{C}(S)} \overline{\mathscr{F}}_\mu(S - \Lambda);$$

then we have

$$\overline{\mathscr{F}}^0_\mu \supset \mathscr{F}^0_\mu = \{\Omega \in \mathscr{F}(S) : \mu(\Omega) = 0 \text{ or } 1\}$$

and for reasons that should become apparent later we will say that μ has *short range correlations* if $\overline{\mathscr{F}}^0_\mu = \mathscr{F}^0_\mu$. We will show that if $V \in H(S)$ and $\mu \in \mathscr{G}_V$ then μ is an extreme point of $\mathscr{G}_V$ if and only if μ has short range correlations. The following results are easily verified (and are left to the reader):

PROPOSITION 11.2 If $\mu \in \mathscr{S}(S)$ then μ has short range correlations if and only if for some (and thus all) p with $1 \leqslant p \leqslant \infty$ we have I_μ^p consists exactly of the constant functions.

LEMMA 11.1 Let D be a non-empty, convex subset of $\mathscr{S}(S)$, and let $\mu \in D$. Then μ is not an extreme point of D if and only if there exists $h \in L_\mu^\infty$, with h not a constant, and such that $h\mu \in D$.

We now need the following definition: if $f \in M(\mathscr{P}(S), \mathscr{F}(S))$ and $\Lambda \in \mathscr{C}(S)$ then we define f_Λ by

$$f_\Lambda(A, X) = f(A \cup X) \quad \text{for} \quad A \subset \Lambda, X \in \mathscr{P}(S - \Lambda).$$

Thus for $A \subset \Lambda$ we have $f_\Lambda(A, \cdot) \in M(\mathscr{P}(S-\Lambda), \mathscr{F}(S-\Lambda))$. If $\mu \in \mathscr{S}(S)$ and $h \in M(\mathscr{P}(S), \mathscr{F}(S), \mu)$ then we can define $h_\Lambda(A, X)$ in the same way and it is easy to check that $h_\Lambda(A, \cdot)$ is a well-defined element of $M(\mathscr{P}(S-\Lambda), \mathscr{F}(S-\Lambda), r_{S-\Lambda}^A(\mu))$ (i.e. if

$$f, g \in M(\mathscr{P}(S), \mathscr{F}(S))$$

with $f = g$ μ-a.e. then $f_\Lambda(A, \cdot) = g_\Lambda(A, \cdot)\, r_{S-\Lambda}^A(\mu)$-a.e.). Let $h \in L_\mu^1$ with $\mu \in \mathscr{S}(S)$; then it is clear that

$$r_{S-\Lambda}^A(h\mu) = h_\Lambda(A, \cdot)\, r_{S-\Lambda}^A(\mu).$$

PROPOSITION 11.3 Let $V \in H(S)$ and $\mu \in \mathscr{G}_V$; let $h \in L_\mu^\infty$ with $h \geqslant 0$ and $\int h \,\mathrm{d}\mu = 1$. Then $h\mu \in \mathscr{G}_V$ if and only if $h \in I_\mu^\infty$.

Proof Let $\{f^\Lambda\}_{\Lambda \in \mathscr{C}(S)}$ be the local specification corresponding to the potential V; let $A \subset \Lambda \in \mathscr{C}(S)$. Then since f^Λ is strictly positive and $r_{S-\Lambda}^A(\mu) = f^\Lambda(A, \cdot)\, r_{S-\Lambda}(\mu)$ we have that $r_{S-\Lambda}^A(\mu)$ and $r_{S-\Lambda}(\mu)$ have the same null sets. Thus if $g \in M(\mathscr{P}(S), \mathscr{F}(S), \mu)$ then we can consider $g_\Lambda(A, \cdot)$ as an element of

$$M(\mathscr{P}(S-\Lambda), \mathscr{F}(S-\Lambda), r_{S-\Lambda}(\mu)).$$

Let $h \in L_\mu^\infty$ with $h \geqslant 0$ and $\int h \,\mathrm{d}\mu = 1$. (The assumptions that $h \geqslant 0$ and $\int h \,\mathrm{d}\mu = 1$ are only needed to ensure that $h\mu \in \mathscr{S}(S)$.) Suppose that $h\mu \in \mathscr{G}_V$; take $A \subset \Lambda \in \mathscr{C}(S)$. Then we have

$$r_{S-\Lambda}^A(h\mu) = f^\Lambda(A, \cdot)\, r_{S-\Lambda}(h\mu),$$

i.e.

$$h_\Lambda(A, \cdot)\, r_{S-\Lambda}^A(\mu) = f^\Lambda(A, \cdot)\, r_{S-\Lambda}(h\mu).$$

But $r^A_{S-\Lambda}(\mu) = f^\Lambda(A,\cdot)\, r_{S-\Lambda}(\mu)$ and thus

$$f^\Lambda(A,\cdot)\, r_{S-\Lambda}(h\mu) = f^\Lambda(A,\cdot)\, h_\Lambda(A,\cdot)\, r_{S-\Lambda}(\mu).$$

Therefore as $f^\Lambda(A,\cdot)$ is strictly positive we have

$$r_{S-\Lambda}(h\mu) = h_\Lambda(A,\cdot)\, r_{S-\Lambda}(\mu).$$

This holds for any $A \subset \Lambda$ and thus

$$h_\Lambda(A,\cdot)\, r_{S-\Lambda}(\mu) = h_\Lambda(B,\cdot)\, r_{S-\Lambda}(\mu) \quad \text{for all} \quad A, B \subset \Lambda.$$

Therefore $h_\Lambda(A,\cdot) = h_\Lambda(B,\cdot)$ for all $A, B \subset \Lambda$ (considered as elements of $L^\infty(\mathscr{P}(S-\Lambda), \mathscr{F}(S-\Lambda), r_{S-\Lambda}(\mu))$. It is easy to check this implies that $h \in L^\infty_\mu(S-\Lambda)$ and since this holds for all $\Lambda \in \mathscr{C}(S)$ we must have $h \in I^\infty_\mu$. Conversely, if $h \in I^\infty_\mu$ then

$$h_\Lambda(A,\cdot) = h_\Lambda(\varnothing,\cdot) \quad \text{for all} \quad A \subset \Lambda$$

and thus

$$\begin{aligned} f^\Lambda(A,\cdot)\, r_{S-\Lambda}(h\mu) &= f^\Lambda(A,\cdot) \sum_{B\subset\Lambda} r^B_{S-\Lambda}(h\mu) \\ &= f^\Lambda(A,\cdot) \sum_{B\subset\Lambda} h_\Lambda(B,\cdot)\, r^B_{S-\Lambda}(\mu) \\ &= f^\Lambda(A,\cdot) \sum_{B\subset\Lambda} h_\Lambda(\varnothing,\cdot)\, r^B_{S-\Lambda}(\mu) = f^\Lambda(A,\cdot)\, h_\Lambda(\varnothing,\cdot)\, r_{S-\Lambda}(\mu) \\ &= h_\Lambda(\varnothing,\cdot)\, r^A_{S-\Lambda}(\mu) = h_\Lambda(A,\cdot)\, r^A_{S-\Lambda}(\mu) = r^A_{S-\Lambda}(h\mu). \end{aligned}$$

Therefore $$r^A_{S-\Lambda}(h\mu) = f^\Lambda(A,\cdot)\, r_{S-\Lambda}(h\mu),$$

and hence $$h\mu \in \mathscr{G}_V. \square$$

From Propositions 11.2, 11.3 and Lemma 11.1 we immediately get:

Theorem 11.1 Let $V \in H(S)$ and $\mu \in \mathscr{G}_V$. Then μ is an extreme point of $\mathscr{G}_V$ if and only if μ has short range correlations.

We will now investigate the properties of states having short range correlations.

Proposition 11.4 The following are equivalent for $\mu \in \mathscr{S}(S)$:

(i) μ has short range correlations;

(ii) given any $A \subset B \in \mathscr{C}(S)$ and $\epsilon > 0$ then there exists $\Lambda \in \mathscr{C}(S)$ such that

$$|\mu(\Omega \cap [A, B]) - \mu(\Omega)\,\mu([A, B])| \leqslant \epsilon$$

whenever $\Omega \in \overline{\mathscr{F}}(S - \Lambda)$;

(iii) given any $f \in L^1_\mu$ then there exists $\Lambda \in \mathscr{C}(S)$ such that

$$\left|\int fg\,\mathrm{d}\mu - \int f\,\mathrm{d}\mu \int g\,\mathrm{d}\mu\right| \leqslant \|g\|_\infty$$

whenever $g \in L^\infty_\mu(S - \Lambda)$;

(iv) if p and q are such that $1 < p \leqslant \infty$ and $1/p + 1/q = 1$ then given $f \in L^1_\mu$ there exists $\Lambda \in \mathscr{C}(S)$ such that

$$\left|\int fg\,\mathrm{d}\mu - \int f\,\mathrm{d}\mu \int g\,\mathrm{d}\mu\right| \leqslant \|g\|_p$$

whenever $g \in L^p_\mu(S - \Lambda)$.

Proof (i) $\Rightarrow$ (iii): Suppose that (iii) does not hold. Then there exists $f \in L^1_\mu$ and $\Lambda_n \in \mathscr{C}(S)$ with $\Lambda_n \uparrow S$ and there exist

$$g_n \in L^\infty_\mu(S - \Lambda_n) \quad \text{with} \quad \|g_n\|_\infty \leqslant 1$$

and
$$\lim_{n \to \infty} \left\{\int fg\,\mathrm{d}\mu - \int f\,\mathrm{d}\mu \int g_n\,\mathrm{d}\mu\right\} \neq 0.$$

Now L^1_μ is separable and $(L^1_\mu)^* = L^\infty_\mu$; thus by taking a subsequence we can assume that there exists $g \in L^\infty_\mu$ such that $g_n \to g$ (in the weak * topology) as $n \to \infty$. But clearly $g \in I^\infty_\mu$ and we have

$$\int fg\,\mathrm{d}\mu - \int f\,\mathrm{d}\mu \int g\,\mathrm{d}\mu = \lim_{n \to \infty} \left\{\int fg_n\,\mathrm{d}\mu - \int f\,\mathrm{d}\mu \int g_n\,\mathrm{d}\mu\right\};$$

hence
$$\int fg\,\mathrm{d}\mu \neq \int f\,\mathrm{d}\mu \int g\,\mathrm{d}\mu.$$

Thus g is not a constant and therefore μ does not have short range correlations.

(iii) $\Rightarrow$ (ii): This is clear.

(ii) $\Rightarrow$ (i): Suppose (ii) holds and let $\Omega \in \overline{\mathcal{F}}^0_\mu$. Note that if $\Lambda \in \mathcal{C}(S)$ then there exists $\Omega' \in \overline{\mathcal{F}}(S-\Lambda)$ such that $\mu(\Omega\Omega') = 0$. Thus if $A \subset B \in \mathcal{C}(S)$ we must have

$$|\mu(\Omega \cap [A,B]) - \mu(\Omega)\,\mu([A,B])| \leqslant \epsilon \quad \text{for all} \quad \epsilon > 0$$

and hence $$\mu(\Omega \cap [A,B]) = \mu(\Omega)\,\mu([A,B]).$$

Since this holds for all finite dimensional cylinders $[A,B]$ we must have either $\mu(\Omega) = 0$ or $\mu(\Omega) = 1$ and thus $\overline{\mathcal{F}}^0_\mu = \mathcal{F}^0_\mu$.

(i) $\Rightarrow$ (iv): This is the same as the proof that (i) $\Rightarrow$ (iii) using the fact that if $1 < p \leqslant \infty$ then L^q_μ is separable and $(L^q_\mu)^* = L^p_\mu$.

(iv) $\Rightarrow$ (ii): This is clear. □

If μ has short range correlations then (ii) says that $[A,B]$ is almost independent of (or uncorrelated with) any event in $\mathcal{F}(S-\Lambda)$. It is from this property that the term 'short range correlations' is derived.

We will continue to look at states having short range correlations by considering the notion of a δ-trivial σ-algebra. Let $\mathcal{B}$ be a sub σ-algebra of $\mathcal{F}(S)$, let $\mu \in \mathcal{S}(S)$ and suppose $0 < \delta \leqslant 1$. We will say that $\mathcal{B}$ is *δ-trivial* with respect to μ if for any $\Omega \in \mathcal{B}$ we either have $\mu(\Omega) = 0$ or $\mu(\Omega) \geqslant \delta$. (This says that up to μ-null sets $\mathcal{B}$ just consists of atoms with μ-measure at least δ.) We will write *trivial* instead of 1-trivial, thus $\mathcal{B}$ is trivial with respect to μ if and only if $\mathcal{B} \subset \mathcal{F}^0_\mu$. Note that if $\mathcal{B}$ is δ-trivial with $\delta > \frac{1}{2}$ then $\mathcal{B}$ is in fact trivial, since if $\Omega \in \mathcal{B}$ with $\mu(\Omega) > 0$ then we have $\mu(\Omega) > \frac{1}{2}$, thus $\mu(\mathcal{P}(S) - \Omega) < \frac{1}{2}$ and so $\mu(\mathcal{P}(S) - \Omega) = 0$, i.e. $\mu(\Omega) = 1$. We will need the following standard fact:

LEMMA 11.2 Let $\mu \in \mathcal{S}(S)$, $\Lambda \in \mathcal{C}(S)$ and $\Omega \in \overline{\mathcal{F}}_\mu(S-\Lambda)$. Then given any $\epsilon > 0$ there exists $B \in \mathcal{C}(S-\Lambda)$ and $A_1, \ldots, A_n \subset B$ such that

$$\mu\left(\Omega \bigcup_{j=1}^{n} [A_j, B]\right) < \epsilon,$$

i.e. $$\mu\left(\left(\Omega - \bigcup_{j=1}^{n} [A_j, B]\right) - \left(\bigcup_{j=1}^{n} [A_j, B] - \Omega\right)\right) < \epsilon.$$

Proof We leave this to the reader. (The proof follows easily from the fact that $\{[A,B]: A \subset B \in \mathscr{C}(S-\Lambda)\}$ is a basis of open sets for the topology on $\mathscr{P}(S-\Lambda)$, and from the fact that $r_{S-\Lambda}(\mu)$ is a regular measure on $\mathscr{P}(S-\Lambda)$.) $\square$

PROPOSITION 11.5 Let $\mu \in \mathscr{S}(S)$ and $0 < \delta \leqslant 1$. Suppose that given any $\epsilon > (1-\delta)$ and $A \subset B \in \mathscr{C}(S)$ there exists $\Lambda \in \mathscr{C}(S)$ such that

$$|\mu([A,B] \cap \Omega) - \mu([A,B])\,\mu(\Omega)| \leqslant \epsilon\mu([A,B])$$

whenever $\Omega \in \overline{\mathscr{F}}(S-\Lambda)$. Then $\overline{\mathscr{F}}^0_\mu$ is δ-trivial with respect to μ.

Proof Let $\Omega \in \overline{\mathscr{F}}^0_\mu$; then given any $\Lambda \in \mathscr{C}(S)$ there exists $\Omega' \in \overline{\mathscr{F}}(S-\Lambda)$ such that $\mu(\Omega\Omega') = 0$. Therefore given any $A \subset B \in \mathscr{C}(S)$ we must have

$$|\mu([A,B] \cap \Omega) - \mu([A,B])\,\mu(\Omega)| \leqslant (1-\delta)\,\mu([A,B]).$$

Now given any $\eta > 0$ we have from Lemma 11.2 that there exists $B \in \mathscr{C}(S)$ and $A_1, \ldots, A_n \subset B$ such that

$$\mu\left(\Omega \bigcup_{j=1}^{n} [A_j, B]\right) < \eta;$$

and clearly we can assume that all the A_j are distinct. Thus we have

$$\begin{aligned}
\mu(\Omega)\,(1-\mu(\Omega)) &= \mu(\Omega) - \mu(\Omega)\,\mu(\Omega) \\
&\leqslant \mu\left(\Omega \cap \left(\bigcup_{j=1}^{n} [A_j, B]\right)\right) - \mu\left(\bigcup_{j=1}^{n} [A_j, B]\right)\mu(\Omega) + 2\eta \\
&= 2\eta + \sum_{j=1}^{n} (\mu(\Omega \cap [A_j, B]) - \mu([A_j, B])\,\mu(\Omega)) \\
&\leqslant 2\eta + \sum_{j=1}^{n} |\mu(\Omega \cap [A_j, B]) - \mu([A_j, B])\,\mu(\Omega)| \\
&\leqslant 2\eta + (1-\delta)\sum_{j=1}^{n} \mu([A_j, B]) \\
&= 2\eta + (1-\delta)\,\mu\left(\bigcup_{j=1}^{n} [A_j, B]\right) \\
&\leqslant 2\eta + (1-\delta)\,(\mu(\Omega) + \eta).
\end{aligned}$$

But $\eta > 0$ was arbitrary, therefore

$$\mu(\Omega)(1-\mu(\Omega)) \leqslant \mu(\Omega)(1-\delta).$$

Hence either $\mu(\Omega) = 0$ or $\mu(\Omega) \geqslant \delta$; thus $\overline{\mathscr{F}}^0_\mu$ is δ-trivial. □

Combining Proposition 11.4 and Proposition 11.5 we get:

PROPOSITION 11.6 If $\mu \in \mathscr{S}(S)$ then μ has short range correlations if and only if there exists $\alpha < \frac{1}{2}$ such that given $A \subset B \in \mathscr{C}(S)$ then there exists $\Lambda \in \mathscr{C}(S)$ such that

$$|\mu(\Omega \cap [A,B]) - \mu(\Omega)\mu([A,B])| \leqslant \alpha\mu([A,B])$$

whenever $\Omega \in \overline{\mathscr{F}}(S-\Lambda)$.

Proof If the condition holds, then by Proposition 11.5 we have that $\overline{\mathscr{F}}^0_\mu$ is $(1-\alpha)$-trivial and thus trivial with respect to μ. Hence $\overline{\mathscr{F}}^0_\mu = \mathscr{F}^0_\mu$; i.e. μ has short range correlations. The converse follows immediately from Proposition 11.4. □

If the proof of Proposition 11.5 is re-examined and use is made of Lemma 11.2 then it is quite easy to see that we have the following version of Proposition 11.6: Let $\Lambda_n \in \mathscr{C}(S)$ with $\Lambda_n \uparrow S$, and let $\mu \in \mathscr{S}(S)$. Then μ has short range correlations if and only if there exists $\alpha < \frac{1}{2}$ such that given n and $A \subset \Lambda_n$ then there exists $\Lambda \in \mathscr{C}(S)$ such that

$$\left|\sum_{j=1}^{m} (\mu([E_j \cup A, F \cup \Lambda_n]) - \mu([E_j, F])\mu([A,\Lambda_n]))\right| \leqslant \alpha\mu([A,\Lambda_n])$$

whenever $F \in \mathscr{C}(S-\Lambda)$ and $E_1, \ldots, E_m \subset F$ with all the E_j distinct. From this we immediately get:

PROPOSITION 11.7 Let $\Lambda_n \in \mathscr{C}(S)$ with $\Lambda_n \uparrow S$, let $\alpha < \frac{1}{2}$ and $\mu \in \mathscr{S}(S)$. Suppose that given any n and $A \subset \Lambda_n$ there exists $\Lambda \in \mathscr{C}(S)$ such that

$$|\mu([E \cup A, F \cup \Lambda_n]) - \mu([E,F])\mu([A,\Lambda_n])| \leqslant \alpha\mu([E,F])\mu([A,\Lambda_n])$$

whenever $E \subset F \in \mathscr{C}(S-\Lambda)$. Then μ has short range correlations.

Let $V \in H(S)$ and suppose we could show that every $\mu \in \mathscr{G}_V$ has short range correlations (for example by verifying that any

$\mu \in \mathscr{G}_V$ satisfies the hypotheses of Proposition 11.7). Then by Theorem 11.1 $\mathscr{G}_V$ can only consist of one element, i.e. phase transition does not occur for V. As an example let $S = \mathbb{Z}$ and let V be a nearest neighbour pair potential on $\mathbb{Z}$ with associated bilinear form U. Let

$$\beta_1 = \sup_{|x-y|=1} U(x,y), \quad \beta_2 = \inf_{|x-y|=1} U(x,y),$$

and suppose that $\beta_1 - \beta_2 < \frac{1}{6}\log\frac{3}{2}$. We leave it as an exercise for the reader to show (using Proposition 5.5) that the hypotheses of Proposition 11.7 must hold for any $\mu \in \mathscr{G}_V$ and thus phase transition cannot occur for V.

NOTES The characterization of the extreme points of $\mathscr{G}_V$ and the properties of states with short range correlations are adapted from Lanford and Ruelle (1969). For results on δ-trivial σ-algebras see Iosifescu (1972).

Appendix: *The Lee–Yang circle theorem revisited*

We give here another proof of the Lee–Yang circle theorem, due to Asano, which in some ways is more natural than the proof given in Chapter 9. To refresh the reader's memory we restate the theorem:

THEOREM A.1 (*Lee–Yang circle theorem*) Let $\Lambda = \{1, \ldots, n\}$ and let $(B_{ij})_{1 \leqslant i,\, j \leqslant n}$ be a symmetric $n \times n$ matrix with $0 < B_{ij} \leqslant 1$ for all i, j. Define a polynomial $p(z_1, \ldots, z_n)$ in the n complex variables $z_1, \ldots, z_n$ by

$$p(z_1, \ldots, z_n) = \sum_{A \subset \Lambda} z^A C(A),$$

where
$$z^A = \prod_{i \in A} z_i \quad \text{if} \quad A \neq \varnothing, \quad z^{\varnothing} = 1,$$

$$C(A) = \prod_{i \in A} \prod_{j \in \Lambda - A} B_{ij} \quad \text{if} \quad A \subset \Lambda \quad \text{with} \quad \varnothing \neq A \neq \Lambda,$$

and $C(\varnothing) = C(\Lambda) = 1$.

Let $\xi_1, \ldots, \xi_n \in \mathbb{C}$ with $|\xi_i| \geqslant 1$ for $i = 1, \ldots, n-1$, and

$$p(\xi_1, \ldots, \xi_n) = 0.$$

Then $|\xi_n| \leqslant 1$.

Note that since we have

$$p(z_1^{-1}, \ldots, z_n^{-1}) = z_1^{-1} \ldots z_n^{-1} p(z_1, \ldots, z_n),$$

we can change the theorem to: let $\xi_1, \ldots, \xi_n \in \mathbb{C}$ with $|\xi_i| \leqslant 1$ for $i = 1, \ldots, n-1$, and $p(\xi_1, \ldots, \xi_n) = 0$; then $|\xi_n| \geqslant 1$.

Recall from Chapter 7 that for $A \subset \Lambda$ we defined

$$\sigma_A : \mathscr{P}(\Lambda) \rightarrow \{-1, 1\}$$

by
$$\sigma_A(B) = (-1)^{|A \cap B|}.$$

Let us define $J: \mathscr{P}(\Lambda) \to \mathbb{R}$ by

$$J(\{i,j\}) = -\log B_{ij} \quad \text{if} \quad i \neq j,$$
$$J(B) = 0 \quad \text{if} \quad |B| \neq 2.$$

Then for all $B \subset \Lambda$ we have $J(B) \geqslant 0$ and it is easy to check that for any $A \subset \Lambda$ we have

$$\sum_{B \subset \Lambda} \sigma_B(A)\, J(B) = 2 \sum_{i \in A} \sum_{j \in \Lambda - A} \log B_{ij} - \tfrac{1}{2} \sum_{i \in \Lambda} \sum_{j \in \Lambda} \log B_{ij}.$$

Thus an equivalent formulation of the Lee–Yang circle theorem is:

THEOREM A.2 Let $\Lambda = \{1, \ldots, n\}$ and let $J: \mathscr{P}(\Lambda) \to \mathbb{R}$ with $J(A) \geqslant 0$ for all $A \subset \Lambda$ and $J(A) = 0$ unless $|A| = 2$. Define a polynomial $p(z_1, \ldots, z_n)$ in the n complex variables $z_1, \ldots, z_n$ by

$$p(z_1, \ldots, z_n) = \sum_{A \subset \Lambda} z^A \exp\{\sum_{B \subset \Lambda} \sigma_B(A)\, J(B)\}.$$

Let $\xi_1, \ldots, \xi_n \in \mathbb{C}$ and suppose $p(\xi_1, \ldots, \xi_n) = 0$. Then for some i we must have $|\xi_i| \geqslant 1$.

For $i \in \Lambda$ let

$$b_i(J) = \text{number of } j \in \Lambda \text{ such that } J(\{i,j\}) > 0,$$

and let
$$b(J) = \sum_{i \in \Lambda} \max\{b_i(J) - 1, 0\}.$$

$b(J)$ gives a measure of the complexity of the interaction J; we will prove the theorem by induction on $b(J)$. Firstly suppose that $b(J) = 0$; then there exists a partition $\Lambda_1 \cup \Lambda_2 \cup \ldots \cup \Lambda_r$ of Λ such that $|\Lambda_s| = 1$ or 2 and if $J(\{i,j\}) > 0$ then $\{i,j\} = \Lambda_s$ for some s. An easy calculation gives that

$$p(z_1, \ldots, z_n) = \prod_{s=1}^{r} p_s(z_1, \ldots, z_n),$$

where
$$p_s(z_1, \ldots, z_n) = \sum_{A \subset \Lambda_s} z^A \exp\{\sum_{B \subset \Lambda_s} \sigma_B(A)\, J(B)\}.$$

Now if $\Lambda_s = \{i\}$ then $p_s(z_1, \ldots, z_n) = 1 + z_i$, and if $\Lambda_s = \{i,j\}$ then

$$p_s(z_1, \ldots, z_n) = \beta(1 + \alpha z_i + \alpha z_j + z_i z_j),$$

with
$$\alpha = \exp[-2J(\{i,j\})], \quad \beta = \exp J(\{i,j\}).$$

Thus for any s we have that $p_s(\xi_1, \ldots, \xi_n) = 0$ implies that $|\xi_i| \geqslant 1$ for some i. But if $p(\xi_1, \ldots, \xi_n) = 0$ then for some s we must have $p_s(\xi_1, \ldots, \xi_n) = 0$, hence the theorem is true in the case that $b(J) = 0$.

The crucial result to do the induction is the following:

LEMMA A.1 Let F_1, F_2 be closed subsets of $\mathbb{C}$ that do not contain 0. Suppose that the complex polynomial

$$a + bz_1 + cz_2 + dz_1 z_2$$

can only vanish when either $z_1 \in F_1$ or $z_2 \in F_2$. Then $a + dz$ can only vanish when $z \in -F_1F_2$ (where

$$-F_1F_2 = \{-\xi_1\xi_2 : \xi_1 \in F_1, \xi_2 \in F_2\}).$$

Let us omit the proof of the lemma for the moment and complete the proof of the theorem. We suppose that $b(J) = m \geqslant 1$ and that the theorem is true for $b(J) < m$. Take $\{i, j\} \subset \Lambda$ such that $J(\{i, j\}) > 0$ and $b_i(J) \geqslant 2$. (This can be done since $b(J) \geqslant 1$.) Let $i' \notin \Lambda$ and put $\Lambda' = \Lambda \cup \{i'\}$, $\Lambda'' = \Lambda - \{i\}$; define $J' : \mathscr{P}(\Lambda') \to \mathbb{R}$ by

$$J'(B) = J(B) \quad \text{if} \quad B \subset \Lambda \quad \text{and} \quad B \neq \{i, j\},$$

$$J'(\{i', j\}) = J(\{i, j\})$$

$$J'(B) = 0 \quad \text{otherwise.}$$

Then $b(J') = b(J) - 1$. Now let $\xi_k \in \mathbb{C}$ with $|\xi_k| < 1$ for $k \in \Lambda''$, and write

$$\sum_{A \subset \Lambda'} \xi^A \exp\{\sum_{B \subset \Lambda'} \sigma_B(A) J'(B)\} = a + b\xi_i + c\xi_{i'} + d\xi_i\xi_{i'}.$$

Since $b(J') < m$ we have by the induction hypothesis that $a + bz_1 + cz_2 + dz_1z_2$ can only vanish if either $|z_1| \geqslant 1$ or $|z_2| \geqslant 1$. Therefore by the lemma, $a + dz$ can only vanish if $|z| \geqslant 1$. But we have

$$a = \sum_{A \subset \Lambda''} \xi^A \exp\{\sum_{B \subset \Lambda'} \sigma_B(A) J'(B)\} = \sum_{A \subset \Lambda''} \xi^A \exp\{\sum_{B \subset \Lambda} \sigma_B(A) J(B)\}$$

and
$$d = \sum_{A \subset \Lambda''} \xi^A \exp\{\sum_{B \subset \Lambda'} \sigma_B(A \cup i \cup i') J'(B)\}$$

$$= \sum_{A \subset \Lambda''} \xi^A \exp\{\sum_{B \subset \Lambda} \sigma_B(A \cup i) J(B)\}.$$

Thus $$a+d\xi_i = \sum_{A\subset\Lambda} \xi^A \exp\{\sum_{B\subset\Lambda} \sigma_B(A)\,J(B)\}$$

and hence we have completed the induction step.

It now only remains to give a proof of the lemma: Clearly we must have $a \neq 0$ (since $0 \notin F_1, 0 \notin F_2$) and thus we can assume that $d \neq 0$. If $ad-bc = 0$ then

$$a+bz_1+cz_2+dz_1z_2 = d(z_1+c/d)\,(z_2+a/c)$$

and thus $-c/d \in F_1$, $-a/c \in F_2$ and of course $a+zd$ can only vanish if $z = -a/d \in -F_1F_2$. We are therefore left with the case $ad-bc \neq 0$. Let us define linear fractional transformations T_1, T_2 by

$$T_1 z = -\frac{a+bz}{c+dz},$$

$$T_2 z = \frac{a}{dz};$$

we consider T_1 and T_2 as mappings from the Riemann sphere into itself. As $ad-bc \neq 0$, T_1 is non-singular and clearly T_2 is non-singular, thus both T_1 and T_2 are bijections of the Riemann sphere onto itself; note that $T_2^{-1} = T_2$. Define $T = T_1T_2$; then

$$Tz = -\frac{adz+ab}{cdz+ad};$$

thus we have $$ab+adz+ad(Tz)+cdz(Tz) = 0.$$

Therefore as T is a bijection it follows that T is an involution, i.e. $T^2z = z$ for all z. If we let $F_1' = F_1 \cup \{\infty\}, F_2' = F_2 \cup \{\infty\}$, then we cannot have

$$T(F_2') \subsetneqq F_2'$$

(since then we would have

$$F_2' = T^2(F_2') \subset T(F_2') \subsetneqq F_2').$$

Thus $$T(F_2') \cap \overline{\mathbb{C}-F_2} \neq \varnothing.$$

Now take $z \in \mathbb{C}-F_2$ and let w be such that $z = T_1w$. Then

$$a+bw+cz+dzw = 0$$

and as $z \notin F_2$ we have by hypothesis that $w \in F_1$; hence

$$\mathbb{C} - F_2 \subset T_1(F_1)$$

and thus
$$\overline{\mathbb{C} - F_2} \subset \overline{T_1(F_1)} \subset T_1(F_1').$$

Therefore
$$T(F_2') \cap T_1(F_1') \neq \varnothing,$$

i.e.
$$T_1 T_2(F_2') \cap T_1(F_1') \neq \varnothing,$$

and thus
$$T_2(F_2') \cap F_1' \neq \varnothing.$$

Let $w \in T_2(F_2') \cap F_1'$; then $w \in \mathbb{C}$ (i.e. $w \neq \infty$), $w \in F_1$, $a/dw \in F_2$, and hence

$$-\frac{a}{d} = -w\frac{a}{dw} \in -F_1 F_2,$$

which completes the proof of the lemma. □

Using the lemma in the same way as above we can obtain the following generalization of the Lee–Yang circle theorem:

THEOREM A.3 Let Λ be a finite set and let $J: \mathscr{P}(\Lambda) \to \mathbb{R}$. Suppose for each $A \subset \Lambda$ and $x \in A$ we are given a closed subset $M(x, A)$ of $\mathbb{C}$ with $0 \notin M(x, A)$, and such that the polynomial

$$\sum_{X \subset A} z^X \exp[\sigma_X(A) J(A)]$$

does not vanish if $z_x \notin M(x, A)$ for all $x \in A$. Then the polynomial

$$\sum_{A \subset \Lambda} z^A \exp\{\sum_{B \subset \Lambda} \sigma_A(B) J(B)\}$$

does not vanish if

$$z_x \notin -\prod_{x \in A \subset \Lambda} (-M(x, A)) \quad \text{for all} \quad x \in \Lambda.$$

Proof We leave this for the reader. □

In order to use the generalized Lee–Yang circle theorem we have of course to locate the zeros of

$$\sum_{X \subset A} z^X \exp[\sigma_X(A) J(A)].$$

Now it is a simple matter to check that

$$\sum_{X \subset A} z^X \exp[\sigma_X(A)\, J(A)]$$
$$= \cosh J(A) \prod_{x \in A} (1+z_x) + \sinh J(A) \prod_{x \in A} (1-z_x),$$

and hence $$\sum_{X \subset A} z^X \exp[\sigma_X(A)\, J(A)] = 0$$

if and only if $$\prod_{x \in A} \left(\frac{1+z_x}{1-z_x}\right) = -\tanh J(A).$$

Thus we could take

$$M(x, A) = \left\{z \in \mathbb{C}: \left|\frac{1+z}{1-z}\right| \leqslant r(x, A)\right\}$$

where $0 \leqslant r(x, A) < 1$ and

$$\prod_{x \in A} r(x, A) \geqslant |\tanh J(A)|.$$

Note that if $z \in -M(x, A)$ then

$$|\arg z| \leqslant \sin^{-1}[\tfrac{1}{2} r(x, A)]$$

(since $|(1-z)/(1+z)| = r$ is the equation of a circle which is symmetric about the real axis and crosses the real axis at $(1-r)/(1+r)$ and $(1+r)/(1-r)$). Therefore we have the following result:

PROPOSITION A.1 Let Λ be a finite set and let $J: \mathscr{P}(\Lambda) \to \mathbb{R}$. Suppose for each $A \subset \Lambda$ and $x \in A$ we have $r(x, A)$ with

$$0 \leqslant r(x, A) < 1$$

and $$\prod_{x \in A} r(x, A) \geqslant |\tanh J(A)|.$$

Suppose also that there exists $\alpha > 0$ such that

$$\sum_{x \in A \subset \Lambda} \sin^{-1}\left[\frac{r(x, A)}{2}\right] \leqslant \pi - \alpha \quad \text{for all} \quad x \in \Lambda.$$

Then the polynomial

$$\sum_{A\subset\Lambda} z^A \exp\{\sum_{B\subset\Lambda} \sigma_A(B)\, J(B)\}$$

does not vanish if $|\arg z_x| < \alpha$ for all $x \in \Lambda$.

If we have $J(A) \geqslant 0$ for all $A \subset \Lambda$ then we can use a different approach: since we have

$$\frac{1+z}{1-z} = \frac{1-|z|^2+\mathrm{i}(2\,\mathrm{Im}\,z)}{|1-z|^2}$$

it follows that if $|z| < 1$ then

$$\mathrm{Re}\left(\frac{1+z}{1-z}\right) > 0,$$

and we also have

$$\arg\left(\frac{1+z}{1-z}\right) = \tan^{-1}\left(\frac{2\,\mathrm{Im}\,z}{1-|z|^2}\right).$$

Thus in this case we can take

$$M(x,A) = \begin{cases} \{-1\} & \text{if} \quad J(A) = 0, \\ \{z\colon |z| \geqslant 1\} & \text{if} \quad J(A) > 0 \quad \text{and} \quad |A| = 1 \text{ or } 2, \\ \{z\colon |z| \geqslant 1\} \cup \left\{z\colon \left|\dfrac{2\,\mathrm{Im}\,z}{1-|z|^2}\right| \geqslant \tan\dfrac{\pi}{|A|}\right\} & \text{if} \quad J(A) > 0 \\ \quad \text{and} \quad |A| \geqslant 3. \end{cases}$$

An elementary calculation shows that if $J(A) > 0$ and $|A| \geqslant 3$ then

$$M(x,A) \cap \left\{z\colon |z| < \tan\left(\frac{\pi}{2|A|}\right)\right\} = \varnothing.$$

We therefore have:

PROPOSITION A.2 Let Λ be a finite set and let $J\colon \mathscr{P}(\Lambda) \to \mathbb{R}$ with $J(A) \geqslant 0$ for all $A \subset \Lambda$. For $x \in \Lambda$, $m \geqslant 1$ let

$$n(x,m) = \text{number of } A \subset \Lambda \text{ with } x \in A, \quad |A| = m \quad \text{and} \quad J(A) > 0,$$

and let

$$\alpha = \min_{x\in\Lambda} \prod_{m=1}^{|\Lambda|} \left(\tan\left(\frac{\pi}{2m}\right)\right)^{n(x,m)}.$$

Then the polynomial

$$\sum_{A \subset \Lambda} z^A \exp\{ \sum_{B \subset \Lambda} \sigma_A(B)\, J(B)\}$$

does not vanish if $|z_x| < \alpha$ for all $x \in \Lambda$. (Note that α depends only on where J is not zero, it does not depend on how large J is.)

NOTES The material in the Appendix is taken from Ruelle (1971a), which is based on an idea of Asano (1970).

Bibliography

Note: *Functional Analysis and its Applications* (*Funct. Anal. Appl.*) refers to the English translation of *Funktsional'nyĭ Analiz i Ego Prilozheniya*; *Theory of Probability and its Applications* (*Theory Probab. Appl.*) refers to the English translation of *Teoriya Veroyatnosteĭ i ee Primeneniya.*

ASANO, T. (1970). The rigorous theorems for the Heisenberg ferromagnets. *J. Phys. Soc. Japan*, **29**, 350–359.

AVERINTSEV, M. B. (1970). On a method of describing discrete parameter random fields. *Problemy Peredachi Informatsii*, **6**, 100–109.

BILLINGSLEY, P. (1965). *Ergodic Theory and Information.* John Wiley & Sons, New York.

DOBRUSHIN, R. L. (1965). Existence of a phase transition in the two-dimensional and three-dimensional Ising models. *Theory Probab. Appl.* **10**, 193–213.

DOBRUSHIN, R. L. (1968a). Description of a random field by means of conditional probabilities and the conditions governing its regularity. *Theory Probab. Appl.* **13**, 197–224.

DOBRUSHIN, R. L. (1968b). Gibbsian random fields for lattice systems with pairwise interactions. *Funct. Anal. Appl.* **2**, 292–301.

DOBRUSHIN, R. L. (1968c). The problem of uniqueness of a Gibbsian random field and the problem of phase transition. *Funct. Anal. Appl.* **2**, 302–312.

DOBRUSHIN, R. L. (1969). Gibbsian random fields. The general case. *Funct. Anal. Appl.* **3**, 22–28.

DOBRUSHIN, R. L. (1971). Markov processes with a large number of locally interacting components – existence of the limiting process and its ergodicity. *Problemy Peredachi Informatsii*, **7**, 70–87.

DOBRUSHIN, R. L., PATETSKI-SHAPIRO, I. I. & VASILEV, N. B. (1969). Markov processes in an infinite product of discrete spaces. *Soviet–Japanese Symposium in Probability Theory, Khavarovsk.*

DOOB, J. L. (1953). *Stochastic Processes.* John Wiley & Sons, New York.

FORTUIN, C. M., KASTELYN, P. W. & GINIBRE, J. (1971). Correlation inequalities on some partially ordered sets. *Comm. Math. Phys.* **22**, 89–103.

GIBBS, W. (1902). *Elementary Principles of Statistical Mechanics.* Yale University Press.

GINIBRE, J. (1970). General formulation of Griffiths inequalities. *Comm. Math. Phys.* **16**, 310–328.

GRIFFITHS, R. B. (1964). Peierls' proof of spontaneous magnetization in two-dimensional Ising ferromagnets. *Phys. Rev.* A **136**, 437–438.

GRIFFITHS, R. B. (1967). Correlations in Ising ferromagnets. *J. Math. Phys.* **8**, 478–489.

GRIMMETT, G. R. (1973). A theorem about random fields. *Bull. London. Math. Soc.* **5**, 81–84.

HAMMERSLEY, J. M. & CLIFFORD, P. (1971). Markov fields on finite graphs and lattices, unpublished.

HARRIS, T. E. (1972). Nearest neighbor Markov interaction processes on multidimensional lattices. *Advan. Math.* **9**, 66–89.

HOLLEY, R. (1970). A class of interactions in an infinite particle system. *Advan. Math.* **5**, 291–309.

HOLLEY, R. (1971). Free energy in a Markovian model of a lattice spin system. *Comm. Math. Phys.* **23**, 87–99.

HOLLEY, R. (1973a). Pressure and Helmholtz free energy in a dynamic model of a lattice gas. Proceedings of the Sixth Berkeley Symposium on Probability and Mathematical Statistics, **3**, 565–578.

HOLLEY, R. (1973b). Markovian interaction processes with finite range interactions. *Ann. Math. Statist.* **43**, 1961–1967.

HOLLEY, R. (1973c). Some remarks on the FKG inequalities, to appear.

IOSIFESCU, M. (1972). On finite tail σ-algebras. *Z. Wahrs. verw. Gebiete*, **24**, 159–166.

KELLY, D. G. & SHERMAN, S. (1968). General Griffiths inequalities on correlations in Ising ferromagnets. *J. Math. Phys.* **9**, 466–484.

LANFORD, O. E. & RUELLE, D. (1969). Observables at infinity and states with short range correlations in statistical mechanics. *Comm. Math. Phys.* **13**, 194–215.

LEBOWITZ, J. L. & MARTIN-LÖF, A. (1972). On the uniqueness of the equilibrium state for Ising spin systems. *Comm. Math. Phys.* **25**, 276–282.

LIGGETT, T. M. (1972). Existence theorems for infinite particle systems. *Trans. Am. Math. Soc.* **165**, 471–481.

LIGGETT, T. M. (1973). An infinite particle system with zero range interactions. *Ann. Probab.* **1**, 240–253.

MAYER, J. E. (1947). Integral equations between distribution functions of molecules. *J. Chem. Phys.* **15**, 187–201.

MINLOS, R. A. (1967a). Limiting Gibbs distribution. *Funct. Anal. Appl.* **1**, 140–150.

MINLOS, R. A. (1967b). Regularity of the Gibbs limit distribution. *Funct. Anal. Appl.* **1**, 206–217.

PEIERLS, R. E. (1936). On Ising's ferromagnet model. *Proc. Camb. Phil. Soc.* **32**, 477–481.

PRESTON, C. J. (1973). Generalized Gibbs states and Markov random fields. *Advan. Appl. Probab.* **5**, 242–261.

RUELLE, D. (1967a). States of classical statistical mechanics. *J. Math. Phys.* **8**, 1657–1668.

RUELLE, D. (1967b). A variational formulation of equilibrium statistical mechanics and the Gibbs phase rule. *Comm. Math. Phys.* **5**, 324–329.

RUELLE, D. (1969). *Statistical Mechanics*. Benjamin, New York.

RUELLE, D. (1971a). Extension of the Lee–Yang circle theorem. *Phys. Rev. Lett.* **26**, 303–304.

RUELLE, D. (1971b). On the use of 'small external fields' in the problem of symmetry breakdown in statistical mechanics. *Ann. Phys.* **69**, 364–374.

SHERMAN, S. (1969). Cosets and ferromagnetic correlation inequalities. *Comm. Math. Phys.* **14**, 1–4.

SHERMAN, S. (1973). Markov random fields and Gibbs random fields. *Israel J. Math.* **14**, 92–103.

SPITZER, F. (1970). Interaction of Markov processes. *Advan. Math.* **5**, 246–290.

SPITZER, F. (1971a). Markov random fields and Gibbs ensembles. *Ann. Math. Monthly*, **78**, 142–154.

SPITZER, F. (1971b). Random fields and interacting particle systems. Lectures given to the 1971 M.A.A. Summer Seminar, Math. Assoc. Am.

SUOMELA, P. (1972). Factorings of finite dimensional distributions. *Commentationes Physico-Mathematicae*, **42**, 1–13.

YANG, C. N. & LEE, T. D. (1952). Statistical theory of equations of state and phase transitions. *Phys. Rev.* **87**, 404–409.

Index